MOLECULAR BIOLOGY

Different Facets

MOLECULAR BIOLOGY

Different Facets

Anjali Priyadarshini, PhD
Prerna Pandey, PhD

Apple Academic Press Inc.	Apple Academic Press Inc.
3333 Mistwell Crescent	9 Spinnaker Way
Oakville, ON L6L 0A2 Canada	Waretown, NJ 08758 USA

© 2018 by Apple Academic Press, Inc.

First issued in paperback 2021

Exclusive worldwide distribution by CRC Press, a member of Taylor & Francis Group
No claim to original U.S. Government works

ISBN 13: 978-1-77-463136-2 (pbk)
ISBN 13: 978-1-77-188641-3 (hbk)

Library and Archives Canada Cataloguing in Publication

Priyadarshini, Anjali, author
Molecular biology : different facets / Anjali Priyadarshini, PhD, Prerna Pandey, PhD.
Includes bibliographical references and index.
Issued in print and electronic formats.
ISBN 978-1-77188-641-3 (hardcover).--ISBN 978-1-315-09927-9 (PDF)
1. Molecular biology. I. Pandey, Prerna, author II. Title.

| QH506.P75 2018 | 572.8 | C2018-901454-7 | C2018-901455-5 |

Library of Congress Cataloging-in-Publication Data

Names: Priyadarshini, Anjali, author. | Pandey, Prerna, author.
Title: Molecular biology : different facets / Anjali Priyadarshini, Prerna Pandey.
Description: Toronto ; New Jersey : Apple Academic Press, 2018. | Includes bibliographical references and index.
Identifiers: LCCN 2018008546 (print) | LCCN 2018009409 (ebook) | ISBN 9781315099279 (ebook) | ISBN 9781771886413 (hardcover : alk. paper) Subjects: | MESH: Biochemical Phenomena | Genetic Phenomena | Genetic Techniques | Molecular Biology Classification: LCC QH390 (ebook) | LCC QH390 (print) | NLM QU 34 | DDC 572.8/38--dc23
LC record available at https://lccn.loc.gov/2018008546

Apple Academic Press also publishes its books in a variety of electronic formats. Some content that appears in print may not be available in electronic format. For information about Apple Academic Press products, visit our website at **www.appleacademicpress.com** and the CRC Press website at **www.crcpress.com**

ABOUT THE AUTHORS

Anjali Priyadarshini, PhD

Anjali Priyadarshini, PhD, is an Assistant Professor at Delhi University, India. Dr. Priyadarshini is a Council of Scientific and Industrial Research (CSIR) Government of India awardee. Her field of research and interest includes biotechnology and nanotechnology. Dr. Priyadarshini has published papers in peer-reviewed journals in the biomedical field.

Prerna Pandey, PhD

Prerna Pandey, PhD, is a biotechnologist with several years of wet lab research experience. She has worked at the International Center for Genetic Engineering and Biotechnology, New Delhi, India. Her PhD research involved isolation and molecular characterization of Geminiviruses, genome sequencing, gene annotation, and gene silencing using the RNA interference technology. She has also worked at Transasia Biomedicals and Advance Enzyme Technologies as a scientist. Dr. Pandey has published papers in peer-reviewed journals in the field and has submitted a number of annotated Geminiviral genome sequences to GenBank, including two novel ones. She has also completed editing and proofreading courses from the Society for Editors and Proofreaders (SfEP) and now works as a freelance scientific writer and editor.

Author Details:

Dr. Anjali Priyadarshini, MSc, PhD
Recipient of CSIR: JRF, SRF.
Address: 74 B, Ayodhya Enclave, Sector 13 Rohini, New Delhi 85, India.
Affiliation: Assistant Professor, SRM University, Sonipat, Haryana, India.
E-mail: anjalipriyadarshini1@gmail.com; anjali0419@yahoo.co.in

Dr. Prerna Pandey, MSc, PhD
Address: B 1403, Jasper, Hiranandani Estate, Thane 400607, Maharashtra, India.
Affiliation: Freelance scientific writer and editor.
Phone: +919167932133.
E-mail: prernapandey@gmail.com; prernapandey@hotmail.com

CONTENTS

LIST OF ABBREVIATIONS

AAV	adeno-associated virus
ABA	abscisic acid
ACC	1-aminocyclopropane-1-carboxylate
aCGH	array comparative genomic hybridization
AD	activation domain
AdV	adenovirus
AFLP	amplified fragment length polymorphism
AOH	absence of heterozygosity
ARMS	amplification refractory mutation system
ARS	autonomous replicating sequences
BAC	bacterial artificial chromosome
BD	DNA-binding domain
cAMP	cyclic adenosine mono phosphate
CAP	catabolite activator protein
Cas	CRISPR-associated
Cas9	CRISPR-associated proteins
CBF	C-repeat binding factors
CBLs	calcineurin B-like proteins
CEN	centromeres
CIAP	calf-intestinal alkaline phosphatase
CIPKs	CBL-interacting protein kinases
CKIs	Cdk inhibitors
co-IP	co-immunoprecipitation
CRISPR	clustered regularly interspaced short palindromic repeats
ddNTPs	dideoxynucleotides
DGGE	denaturing gradient gel electrophoresis
DHPLC	denaturing high-performance liquid chromatography
DNA	deoxyribonucleic acid
dNTP	deoxyribonucleotides
DREB	dehydration-responsive element-binding proteins
dsRNA	double-stranded RNA
EMS	ethyl methanesulfonate
ER	endoplasmic reticulum

EST	expressed sequence tag
ET	ethylene
F-SSCP	fluorescent SSCP
FISH	fluorescent in situ hybridization
GE	genetically engineered
GEAC	Genetic Engineering Approval Committee
GGE	gradient gel electrophoresis
GINA	Genetic Information Non-discrimination Act
GM	genetically modified
gRNA	guide RNA
GST	glutathione S-transferase
GWAS	genome-wide association studies
HAD	heteroduplex analyses
HBV	hepatitis B virus
HCN	hydrogen cyanide
HDR	homology-directed repair
Hfr	high frequency recombination
HPV	human papilloma virus
HR	homologous recombination
IAA	indole-3-acetic acid
IF	immunofluorescence
IHF	integration host factor
IP	immunoprecipitation
IRES	internal ribosome sites
IRGSP	International Rice Genome Sequencing Project
IS	insertion sequence
ISO	allele-specific oligonucleotide
IVF	in vitro fertilization
IVSP	in vitro synthesized protein assay
JA	jasmonic acid
KSI	ketosteroid isomerase
LOH	loss of heterozygosity
LVs	lentivirus
MAPH	multiplex amplifiable probe hybridization
MAPKs	mitogen-activated protein kinases
MBP	maltose-binding protein
MCS	multiple cloning site
miRNAs	microRNA

MLPA	multiplex ligation-dependent probe amplification
mRNA	messenger RNA
mRNP	mRNA protein
NGS	next-generation sequencing
NHEJ	nonhomologous end joining
NMPs	nucleoside monophosphates
NPR	nodule promoting rhizobacteria
NTPs	nucleoside triphosphates
OLA	oligonucleotide ligation assay
PAM	protospacer adjacent motif
PCR	polymerase chain reaction
PEG	polyethylene glycol
PGD	pre-implantation genetic diagnosis
PGPR	plant growth promoting rhizobacteria
PHPR	plant health promoting rhizobacteria
PM	personalized medicine
PMF	plant molecular farming
PPi	pyrophosphate
PPV	produced Plum pox virus
PR	plant resistance
pRNA	primer RNA
PTGS	posttranscriptional gene silencing
PTM	posttranslational modification
PTT	protein truncation test
QTL	quantitative trait locus
RBS	ribosome binding site
RCM	rolling circle mechanism
RFLP	restriction fragment length polymorphisms
RISC	RNA-induced silencing complex
RNA	ribonucleic acid
RNAP	RNA polymerase
RNP	ribonucleoprotein
ROS	reactive oxygen species
rRNA	ribosomal RNA
SA	salicylic acid
sgRNA	single-guide RNA
shRNAs	short hairpin RNAs
SINEs	short interspersed nuclear elements

siRNAs	small interfering RNAs
SNPs	single nucleotide polymorphisms
SSB	single-stranded binding protein
SSCP	single-strand conformation polymorphism
SSO	single-strand origin
SSRs	simple sequence repeats
STPs	signal transduction pathways
SUMO	small ubiquitin related modifier
TBP	TATA-binding protein
TEV	tobacco etch virus
TFs	transcription factors
TGGE	temperature gradient gel electrophoresis
TM	theta mechanism
TMV	tobacco mosaic virus
tPA	tissue plasminogen activator
tracrRNA	trans-activating crRNA
Trp	tryptophan
TrxA	thioredoxin A
TTGE	temporal temperature gradient gel electrophoresis
UPD	uniparental isodisomy
UV	ultraviolet
VEGF	vascular endothelial growth factor
VIGS	virus-induced gene silencing
WCR	Western corn rootworm
YAC	yeast artificial chromosome
YCp	yeast centromeric plasmid

PREFACE

The book *Molecular Biology: Different Facets* includes a comprehensive description of the basic tenets of molecular biology, from mechanisms to their elaborate role in gene regulation. The initial sections describe the history of genetics and molecular biology. The book highlights the significance of the molecular approaches for all biological processes in both simple and complex cells. The text also incorporates the most recent references and has been written for students as well as for teachers of molecular biology, molecular genetics, or biochemistry. The authors have described experimental approaches wherever necessary to explore the evidence that led to the development of important concepts and hypotheses that led to significant advances in molecular biology. The book is divided into six chapters; the initial topics cover basic information that help in understanding the advanced topics covered later in the chapters.

INTRODUCTION

The field of molecular biology has taken several giant steps that have a broad range of applications. Various aspects of life came to be known by the scientific pursuits done in molecular biology. Molecular biology has been found to have very significant use in disease diagnostics and therapeutics. These outcomes have been initiated from the discovery of the structure of DNA. This discovery led us to deciphering of the genetic code and its outcome. A very important tenet that forms the core of life processes is the central dogma of life, which was established when the genetic code was decoded. The various aspects of the expression of this genetic code opened up a wide arena of its role played in various biochemical processes. Once the genetic basis of multiple diseases was established, the path was paved for its diagnostic and therapeutic use. Not only its role is there in disease but also aids in regulation with the help of regulatory RNA. All this goes in it tandem with the aid of various other branches of science such as microbiology, genetics, biochemistry, to name a few. So much is known and still much is left to be known. This is a vast field of micromolecules that affects our well-being so much.

The molecular mechanisms underlying the various processes taking place in a cell such as replication, transcription, processing of RNA, and translation offer various new avenues for research with potential in understanding the complex system of life. The study of the cell cycle and various intricacies in its control could serve as checkpoints in understanding diseases and potential drug targets.

Several processes in bacteria like conjugation, transformation, and transduction ushered in an era of excitement and research. The understanding of "simple" bacterial molecular biology changed the paradigm of viewing these microscopic cells with all their intricacies. These molecular mechanisms opened up new avenues as their potential in mapping genes and related genetic studies and their use in recombinant DNA technology, which has and is revolutionizing science. The same statement may be extended to another kingdom of fungi that have their own molecular pathways.

The molecular aspects of the so-called threshold organisms, such as viruses, show immense complexities and elaborately sophisticated pathways.

This book covers aspects of molecular biology in brief, as each topic is an ocean to be delved into. Nevertheless, though each of the topics is presented as "tips of the icebergs," they have been presented and elucidated in concise terms with relevant research. The book serves to ignite the minds of students and academicians to pursue research or just serve as reading material.

CHAPTER 1

CELL

CONTENTS

ABSTRACT

The volume of research that has gone into study and advancement of cell biology requires it to be dealt with independently as a subject. Cell biology deals with the cellular organization of various life forms including prokaryotes, eukaryotes as well as the cellular forms such as viruses. The variations within a group lead to its classification which is very much dependent on the surrounding environment as well. Continuity of the living forms requires division of cell and formation of daughter cells within a generation or to form cells involved in formation of progeny for next generation. This chapter is an attempt to give a panoramic view of the entire process and the subject.

1.1   INTRODUCTION TO CELL

Human mind has been very curious to know when, why, and how of any event, regarding all the biological processes. Our understanding of all the biological process gained a momentum since the invention of simple and complex microscope. First and foremost, breakthrough came with the observations of the cells and the microorganisms by Robert Hooke and Anton von Leeuwenhoek. This led to more detailed study of such micro-scopic structures by more intriguing minds, thus forming foundation for various branches of science such as genetics, biochemistry, molecular biology, cell biology, etc. The subsequent parts of this chapter deals with the various aspects related to aforesaid branches of science. The main focus of this chapter is on the comparative study of animal, plant, and microbial cell apart from dealing with acellular microorganism which is virus.

Molecular biology as the name suggests is the branch of science which deals with the minuscules. All the life processes can now be viewed and analyzed at the microscopic level. This has been made possible by cumu-lative efforts of number of scientist and researchers. I begin my work by thanking all the known, very well-known, and the unknown whose pursuit to find answers to many questions has led to our understanding of molec-ular biology.

Now that the world has seen the smallest of the smallest material ranging in the diameter of nanoscale and less, there was a time when discovery of the basic unit of life, that is, a cell was hailed among the biggest discovery of the smallest. This discovery by Robert Hook laid the foundation of the study of the small as the next big thing. His discovery was the culmination of his simple quest to know the basis of cork functioning to hold air in a bottle. He observed a honeycomb like pattern in cork under microscope, which was nothing but empty cell walls of dead plant tissue as we know now. Robert Hook was able to tantalize our understanding along with another great observer and discoverer Anton van Leeuwenhoek, about the yet to be identified and analyzed microscopic world. Robert Hook was a microscopist and Anton van Leeuwenhoek, a Dutch merchant having a very curious bent of mind and disposition. As Robert Hook is credited with the discovery of cell (Fig. 1.1), Anton van Leeuwenhoek (Fig. 1.2) has the distinction of discovering the animalcules or the microscopic organisms, for example, the bacteria.

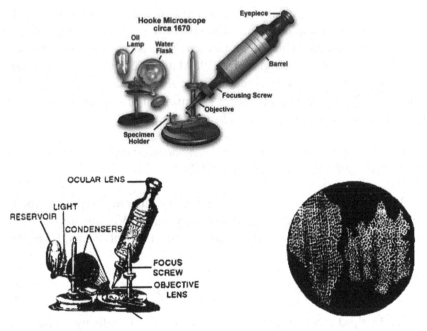

FIGURE 1.1 Robert Hooke's microscope and dead cork cells as seen by him. The cells had hexagonal shape and gave an appearance of bee hive.

FIGURE 1.2   Father of microbiology Anton van Leeuwenhoek.

The next query to be addressed was the ubiquitous nature of cell. If at all, cell could be hailed as the basic unit of life. This was satisfactorily realized by Matthias Schleiden, a German botanist in 1838 and Theodor Schwann, a German zoologist in 1839. Matthias Schleiden in his work concluded that plants were made of cells; similar claims were made by Theodor Schwann regarding cellular organization of animal and proposal of the first two tenets of *The Cell Theory* was made:

1) *All organisms are composed of one or more cells.*
2) *The cell is the structural unit of life.*

Their belief that cells could arise from noncellular material or the spontaneous generation theory was put to rest by Rudolf Virchow in 1855. He was a German pathologist who added to the tenets of cell theory. Later, the third tenet was added.

3) *Cells can arise only by division from a preexisting cell.*

As the third tenet suggests, the formation of new cells by division of preexisting cell has and is being demonstrated in laboratories across the world having basic cell culture facility. Cells can be extracted from plant as well as animals to be "grown" in laboratory. The first human cell to be cultured was obtained from malignant tumor of cervix. It was christened HeLa cells after the donor Henrietta Lacks.

Regarding the first two tenets of cell theory, there are certain properties which are adherent to all the cells irrespective of their origin. They are as follows:

- Response to stimuli: response of a cell against external stimuli with the help of various receptors present on the surface.
- Reproduction: division of cell leading to formation of new daughter cells.
- Acquisition and utilization of energy: photosynthetic and respiratory activities.
- Possession of information in the form of genetic code: following the central dogma of life where the genetic information is transcribed and translated to perform various cellular function.
- Site of various chemical reaction for various life processes: anabolism + catabolism = metabolism.
- Capability to do various mechanical work: transport of various material within a cell or to different locations in the body of multicellular organism.
- Self-regulatory mechanisms: ability to correct malformed genetic information and self-destruction of cells which are beyond repair (apoptosis).

The work of these pioneer researchers in basic science laid the foundation of modern day field of molecular biology, which has helped a great deal in various facets of life including healthcare, diagnostics, therapeutics, and prosthetics to name a few.

1.2 ORGANIZATION AND STRUCTURE OF CELLS

All living things fall broadly into one of the two categories:

- prokaryotes
- eukaryotes

pro = means "prior to," *eu* = means "true," and *karyote* = means "nucleus."

The distinction is based on whether or not a cell has a nucleus. Prokaryotic cells do not have nuclei, while eukaryotic cells do. Also, eukaryotic cells have organelles (Fig. 1.3). Although there are many differences between a prokaryotic cell and a eukaryotic cell, there are many similarities too, which point to the fact that they have a common ancestor.

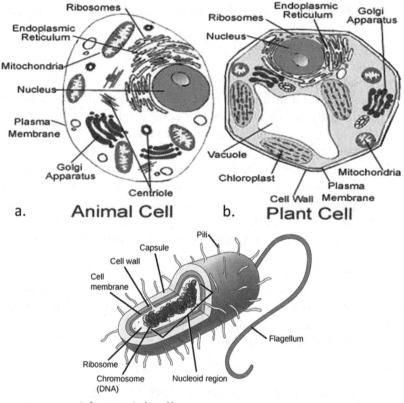

FIGURE 1.3 Schematic diagrams of (a) animal, (b) plant, and (c) bacterial cell. An animal cell lacks cell wall and plastids which is the site for protein synthesis. The plant cell is characterized by presence of large vacuoles. Both plant and animal cell are eukaryotic with a well-defined nucleus. The bacterial cell, which is prokaryotic, lacks a well-defined nucleus, with presence of cell wall made up of peptidoglycan as well as flagella for locomotion.

1.2.1 EARLY EVOLUTION OF CELLS

Contemporary evidence favors the view that all living organisms should be grouped into three lineages (or classes)[1,2,3]:
- archaebacteria
- eubacteria (or just/true bacteria)
- eukaryotes

It is believed (estimated) that evolution of all three lineages occurred approximately 3.5 billion years ago from a common ancestral form called progenote (Table 1.1). Archaea and eubacteria are both prokaryotic because they are single-cell organisms without nucleus. Physiological morphological and ecological diversity has been used to a great extent for classification of prokaryotes. However, it is now accepted that eukaryotic cells are composed of various prokaryotic contributions, for example, mitochondria and chloroplast are considered to be of prokaryotic origin living in a symbiotic association with the eukaryotic cell making the dichotomy artificial.

In contrast to eukaryotes, prokaryotes do not have structural diversity or higher cellular organization such as tissue and organ (*General Microbiology*, 5th edition, Stanier).

1.2.2 ARCHAEBACTERIA

- The word archaea is derived from the word "ancient."
- It is most recently discovered lineage.
- Archaebacteria are similar in shape to bacteria, but genetically they are as distinct from bacteria as they are from eukaryotes.

(Whole genome sequencing of the archaeon *Methanococcus jannaschii* showed 44% similarity to the known genes in eubacteria and 56% of genes that were new to science)

Based on their physiology, archaea can be classified into three subcategories:

a) Methanogens—prokaryotes that produce methane (CH_4)
b) Halophiles—prokaryotes that live at very high concentrations of salt (NaCl)

TABLE 1.1 Classification of Living Things.

Domain	Bacteria	Archaebacteria	Eukaryote			
Kingdom	Eubacteria	Archaebacteria	Protist	Fungi	Plant	Animal
Cell type	Prokaryote	Prokaryote	Eukaryote	Eukaryote	Eukaryote	Eukaryote
Cell structure	Cell wall with peptidoglycan	Cell wall without peptidoglycan	Cell wall of cellulose in some, some have chloroplast	Cell wall with chitin	Cell wall of cellulose and chloroplast	No cell wall of chloroplast
Number of Cells	Unicellular	Unicellular	Most unicellular, some colonial and some multicellular	Most unicellular, some unicellular	Multicellular	Multicellular
Mode of Nutrition	Autotroph or heterotroph	Autotroph or heterotroph	Autotroph or heterotroph	Heterotroph	Autotroph	Heterotroph

c) Thermo-, acido-, or psychrophiles—prokaryotes that live at very high temperatures ($>100°C$) or in acidic environments or at very low temperatures ($10°C$)

1.2.3 EUBACTERIA

- They are ubiquitously single-cell organisms.
- They differ from archaea in chemical content of the cell wall and cell membrane.
- Some important representatives based on morphology, physiology, and ecology are as listed below.
  - **photosynthetic purple and green bacteria (blue-green algae)**—convert the energy of light into chemical energy, but do not produce oxygen
  - **cyanobacteria**—thought to have given rise to eukaryotic chloroplasts; live in fresh water and marine habitats and are a part of a complex microbial community called plankton
  - **spirochetes**—genetically are a distinct group of bacteria; some are pathogens for animals(syphilis, lyme disease, etc.)
  - **spirilla**—live in fresh water and like oxygen; can be pathogenic
  - **myxobacteria** (a group of gliding bacteria)—live in soil or animal dung
  - **lithotrophs**—requires inorganic compounds as sources of energy (this mechanism also exists in archaea). For example, the nitrifying bacteria can convert NH_3 to NO_2 and NO_2 to NO_3; may play an important role in primary production of organic material in nature
  - **pseudomonads and their relatives**—most commonly free-living organisms in soil and water; have flagella
  - **enterics**—can ferment glucose and are present in humans; very well studied; the most important organism is *Escherichia coli* and there are many more

Eubacteria has been differentiated on the basis of cell shape; various shapes have been listed below:

- little balls (cocci)
- medicine capsules (with slimy capsule)

- segmented ribbons (spirochetes)
- little rods (bacillus)

E. coli is one of the very important model and best studied organisms. It is one of the main species of bacteria that live in the lower intestines of warm-blooded animals (including birds and mammals) and are necessary for the proper digestion of food. However, this bacterium may become harmful if it makes itself out of the lower intestines (e.g., dysentery).

1.2.4   VIRUSES

Viruses have been defined as acellular microorganism, which cannot survive outside the host cell and have structure comprising nucleic acid (DNA or RNA) and protein coat. Classification is based on the type of nucleic acid present, which acts as the genetic material (Fig. 1.3d).

Virus

RNA as genome	DNA as genome
Picornavirus (+,C=32, 22-30 nm)	Circovirus (+,icosahedral, 17-22 nm)
Astrovirus (+, C=32, 20-35 nm)	Parvovirus (+ or -, C=12, 18-26 nm)
Calicivirus (+, C=32, 36-39 nm)	Hepadnavirus(+ -,icosahedral,40-48 nm)
Flavivirus (+, icosahedral, 45-50 nm)	Papovavirus (+-, C=72, 45-55 nm)
Togavirus (+, envelope, icosahedral, 70 nm)	Adenovirus (+-, C=262, 75-80 nm)
Coronavirus (+, pleiomorphic, 120-160 nm)	Herpesvirus (+-, C=162, 75-80 nm)
Retrovirus (++, icosahedral, 90-120 nm)	Poxvirus (+-)
Reovirus (+-, C=132, 60-80 nm)	
Bunyavirus (-, 10, 80-120 nm)	
Orthomyxovirus (-, helical, pleiomorphic, 80-120 nm)	
Arenavirus (-, pleiomorphic, 110-120 nm)	
Filovirus (-, helical)	
Rhabdovirus (-, helical, 60-150 nm)	
Paramyxovirus (-, helical, pleiomorphic, 160-300 nm)	

FIGURE 1.3d   Schemes of 21 virus families infecting humans showing a number of distinctive criteria: presence of an envelope or (double-) capsid and internal nucleic acid genome. +, sense strand; antisense strand; ±, dsRNA or DNA; 0, circular DNA; C, number of capsomeres or holes, where known; nm, dimensions of capsid, or envelope when present.

1.2.4.1   GENERAL FEATURES

- They are complexes of nucleic acid (DNA or RNA) encapsulated in a protein coat (capsid) which is in certain cases, outlined by capsule.
- Viruses are obligate cellular parasites.
- The protein coat (capsid) serves to protect the nucleic acid.
- In some instances, it is also surrounded by a membrane.
- Viruses for all types of cells are known (animal virus, plant virus, and bacterial virus).
- Different viruses infect animal, plant, and bacterial cells.
- Viruses infecting bacteria are called bacteriophages ("bacteria eaters").
- Viruses can also cause the lysis (destruction) of cells.
- In some cases, the viral genetic elements may integrate into the host chromosome and become quiescent (also known as lysogeny).
- Some viruses are implicated in transforming cells into a cancerous state, that is, in converting their hosts to an unregulated state of cell division and proliferation.
- Because all viruses are heavily dependent on their host for the production of viral progeny, it was proposed that viruses must have arisen after cells were established in the course of evolution (presumably, the first viruses were fragments of nucleic acid that developed the ability to replicate independently of the chromosome and then acquired the necessary genes enabling protection, autonomy, and transfer between cells).

1.2.5   PROKARYOTIC CELL

Major elements and features of a typical prokaryotic cell (Fig. 1.3c):

- *Cell wall*—a rigid framework of polysaccharide cross-linked by short peptide chains; provides mechanical support, shape, and protection; it is a porous nonselective barrier that allows most small molecules to pass (Fig.1.4).
- *Cell membrane*—45:55% lipid:protein ratio; bilayer; highly selective and controls the entry of most substances into the cell; important proteins are located in the cell membrane.

- *Nucleoid* (DNA)—repository of the cell's genetic information; contains a single tightly coiled DNA molecule.
- *Ribosomes*—sites where proteins are synthesized; consists of a small and a large subunit; a bacterial cell has about 15,000 ribosomes; 35% of a ribosome is protein, the rest is RNA.
- *Storage granules*—granules where polymerized metabolites are stored (e.g., sugars); when needed, the polymers are liberated and degraded by energy-yielding pathways in the cell.
- *Cytosol*—the site of intermediary metabolism (sets of chemical reactions by which cells generate energy and form precursors necessary for biosynthesis of macromolecules essential to cell growth and function.

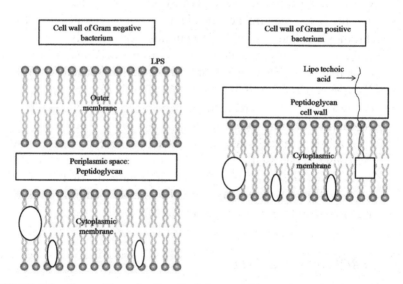

FIGURE 1.4 Composition of cell wall of Gram positive and negative bacteria. Gram positive bacteria have a thick peptidoglycan layer. Gram negative bacteria have thin peptidoglycan layer and there is presence of outer membrane.

1.2.6   EUKARYOTES

The characteristic features of eukaryotic organisms are as follows:

- single-cell or multicell organisms in which each cell contains a nucleus and organelles

- eukaryotes are subdivided into four categories

    - animals—typically divided into vertebrates (e.g., mammals) and invertebrates (e.g., snails)
    - plants—trees, flowers, etc.
    - fungi—sometimes in popular literature considered as plants
    - protists—all other organisms, for example, yeast

1.2.6.1   EUKARYOTIC CELL

Characteristic features of eukaryotic cell are as follows.

- It is much larger in size (1000–10,000 times larger than prokaryotic cells).
- It is much more complex in comparison to a prokaryotic cell.
- Metabolic processes are organized into compartments, with each compartment dedicated to a particular function (enabled by a system of membranes).
- It possesses a nucleus, the repository of cell's genetic material which is distributed among a few or many chromosomes.

1.2.7   ANIMAL CELL

Major elements and features of a typical animal cell (Fig. 1.3a)

- *Extracellular matrix*—a complex coating which is cell specific, serves in cell–cell recognitions and communication, also provides a protective layer.
- *Cell (plasma) membrane*—roughly 50:50% lipid:protein ratio; selectively permeable membrane; contains various systems for influx of extracellular molecules (pumps, channels, and transporters); important proteins are located here.
- *Nucleus*—separated from the cytosol by a double membrane; repository of genetic information—DNA complexed with the basic proteins (histones) to form chromatin fibers, the material from which the chromosomes are made.
- *Nucleolus*—a distinct RNA-rich part of the nucleus where ribosomes are assembled.

- *Mitochondria*—organelles surrounded by two membranes that differ significantly in their protein and lipid composition; mitochondria are power plants of eukaryotic cells where ATP is produced.
- *Golgi apparatus*—involved in packaging and processing of macromolecules for secretion and for delivery of other cellular compartments.
- *Endoplasmic reticulum (ER)*—the ER is an organelle where both membrane proteins and lipids are synthesized.
- *Ribosomes*—organelle is composed of RNA and ribosomal proteins; eukaryotic ribosomes are much larger than prokaryotic ribosomes; attached to ER. Ribosome consists of two subunits that fit together and work as one to translate the mRNA into a polypeptide chain
- *Lysosomes*—function in intracellular digestion of certain materials entering the cell; they also function in the controlled degradation of cellular components.
- *Peroxisomes*—act to oxidize certain nutrients such as amino acids; in doing so, they form potentially toxic hydrogen peroxide and then decompose it by means of the peroxy-cleaving enzyme (protein).
- *Cytoskeleton*—is composed of a network of protein filaments and it determines the shape of the cell and gives it stability; cytoskeleton also mediates internal movements that occur in the cytoplasm, such as migration of organelles and movement of chromosomes during cell division.

1.2.8 PLANT CELL

Major elements and features of a typical plant cell (only what differs from the animal cell has been mentioned) (Fig. 1.3b)

- *Cell wall*—consists of cellulose fibers embedded in a polysaccharide and protein matrix; provides protection from the osmotic and mechanical rupture; channels for fluid circulation and for the cell–cell communication pass through the walls.
- *Chloroplasts*—a unique family of organelles (the plastids) of which, the chloroplast is the prominent example. They are significantly larger than mitochondria and are the site of photosynthesis, the reaction by which light energy is converted to metabolically useful chemical energy in the form of ATP.

- *Mitochondria*—a major source of energy in the dark, involved in ATP production.
- *Vacuole*—a very large vesicle enclosed by a single membrane. They function in transport and storage of nutrients. By accumulating water, the vacuole allows the plant cell to grow dramatically with no increase in cytoplasmic volume.

The study of cell is necessary to lead us to its various subcellular aspects. All this exercise culminates in realizing the basic facts and cellular activity occurring at the molecular level, which forms the basis of study of molecular biology.

1.3 SHARED PROPERTIES OF BIOLOGICAL SYSTEMS

The three processes of replication, transcription, and translation occur in all biological systems, thus making a pertinent point that all the biological systems have some common properties. Moreover all cellular organisms have these three classes of macromolecules: Deoxyribonucleic acid (DNA), Ribonucleic acid (RNA), and proteins—performing well-defined functions. DNA is involved in carrying the coded genetic information except in certain virus, where RNA is the carrier of genetic material. Transmission of this genetic material occurs by mitosis and meiosis in eukaryotes. Mitosis results in the formation of two cells, whereas meiosis, which is also called a reduction division, results in formation of four cells. This genetic information is transcribed by RNA which serves as template for protein synthesis known as translation. Proteins translated include the enzymes which act as biocatalysts for all cellular functions including all the metabolic activity. Cell structure is tightly linked to genetic function. A cell is surrounded by plasma membrane which is actively involved in influx and efflux of materials across it. Eukaryotic cell is characterized by presence of nucleus which is the abode of genetic material, DNA which is arranged in the form of chromatin. The thread like chromatin condenses and forms chromosomes during mitosis and meiosis. The remaining part of cell within plasma membrane is cytoplasm which houses all the cell organelles. Prokaryotes lack the organized nuclear compartmentalization of genetic material. In *E. coli*, the genetic material occurs as a circular thread like DNA molecule residing in large cellular area called nucleoid.

The size of the genome in one of the most well-studied prokaryotes, *E. coli*, is 4.6 million base pairs (approximately 1.1 mm, if cut and stretched out). This exorbitant amount of DNA is packaged into a small bacterial cell having size in micro meter. This is achieved by a very high level of condensation in genetic material. First, DNA is twisted by a process called supercoiling. Supercoiling results in DNA either under-wound (less than one turn of thehelixper 10 base pairs) or over-wound (more than 1 turn per 10 base pairs) from its normal relaxed state. Some proteins are known to be involved in the supercoiling; other proteins and enzymes such as DNA gyrase help in maintaining the supercoiled structure.

Eukaryotes, whose chromosomes each consist of a linear DNA molecule, employ a different type of packing strategy to fit their DNA inside the nucleus. At the most basic level, DNA is wrapped around proteins known as histones to form structures called nucleosomes. The histones are evolutionarily conserved proteins that are rich in basic amino acids and form an octamer. The DNA (which is negatively charged because of the phosphate groups) is wrapped tightly around the histone core. This nucleosome is linked to the next one with the help of a linker DNA. This is also known as the "beads on a string" structure. This is further compacted into a 30nm fiber, which is the diameter of the structure. At the metaphase stage, the chromosomes are at their most compact, approximately 700 nm in width, and are found in association with scaffold proteins (Fig. 1.5).

DNA packaging is supposed to have strong influence on genes' activity. It is also one of the mechanisms of epigenetic control of gene expression.[2,3,5]

1.4 MITOSIS

Mitosis is the mode of asexual reproduction in some single-celled organisms such as protozoa, algae, and fungi. In multicellular higher organisms, mitosis is the mode of growth and development from the zygotic stage as well as wound healing. Mitosis is the mechanism responsible for continuous replacement of cells in various tissues, for example, replacement of epidermal cells and also production of immature blood cells (reticulocytes) which later on shed their nuclei to become mature red blood cells. Loss of cell cycle control results in formation of tumor which are characterized by uncontrolled division and growth of cells. Cell division is a highly coordinated mechanism in which the genetic material is partitioned into daughter cells during nuclear division (karyokinesis) which is followed by cytoplasmic division (cytokinesis) which partitions the whole thing into two halves forming two daughter cells within distinct plasma-membrane. Mitosis has been divided into various division phases, which follow the interphase where only synthesis occurs and no division takes place.

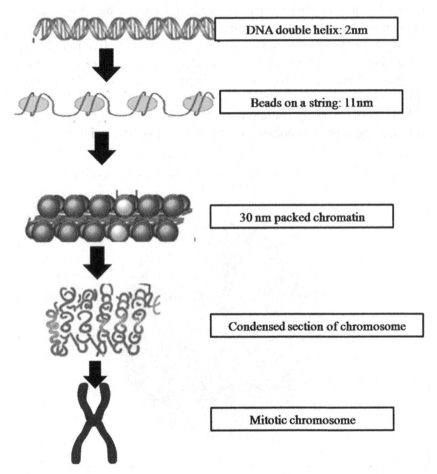

FIGURE 1.5   Packaging and compaction of chromatin into chromosome in eukaryotic cell. DNA in its extended form cannot be packaged within a cell; therefore, it undergoes multiple level of compaction. First level is formation of double helix, which is then bound to histone protein in structure similar to beads on a string. Then, the structure called nucleosome is coiled into chromatin fiber. The chromatin is further condensed and forms chromosome.

1.4.1 INTERPHASE

Interphase is stage between two consecutive cell cycles where active biochemical processes take place, most importantly replication of DNA of each chromosome. The period in which DNA replication takes place is called S or synthesis phase. No DNA synthesis occurs during G1 and G2 phase, also known as gap I and gap II, but this phase is characterized by rigorous metabolic activity, cell growth, and differentiation. End of G2 results in increase in cell volume replication of DNA and initiation of mitosis or M phase. DNA becomes invisible during interphase because DNA loses its compact structure to undergo replication.

Mitosis is a very dynamic phase subdivided into many parts, that is, prophase, metaphase, anaphase, and telophase (Fig.1.6a and b).

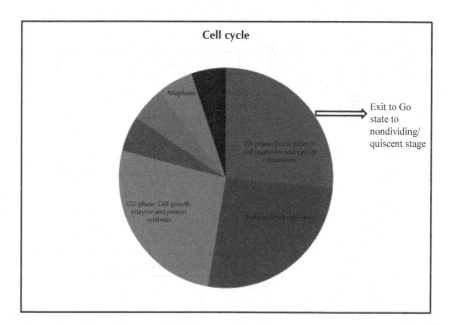

FIGURE 1.6a Cell cycle progression of a normal cell. The phases of cell cycle: the interphase comprising G0, G1, S (synthesis), G2, and M or mitotic phase. Mitosis: which results in the formation of two daughter cells is followed by G1 to start new cell cycle. At this stage, the cell may be unresponsive or nondividing (G0) or commit to complete the cycle through G1, S, and G2.

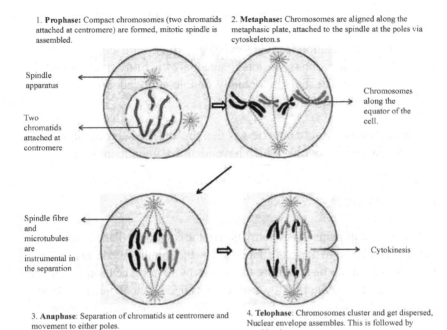

1. **Prophase:** Compact chromosomes (two chromatids attached at centromere) are formed, mitotic spindle is assembled.

2. **Metaphase:** Chromosomes are aligned along the metaphasic plate, attached to the spindle at the poles via cytoskeleton.s

Spindle apparatus

Two chromatids attached at contromere

Chromosomes along the equator of the cell.

Spindle fibre and microtubules are instrumental in the separation

Cytokinesis

3. **Anaphase**: Separation of chromatids at centromere and movement to either poles.

4. **Telophase**: Chromosomes cluster and get dispersed, Nuclear envelope assembles. This is followed by cytokinesis.

FIGURE 1.6b Cell undergoing the mitotic cycle. Prophase is characterized by assembly of mitotic spindle. The metaphase is recognizable by the arrangement of compacted chromatids on the equatorial plane. At anaphase, the separation of sister chromatids toward the opposite poles initiate. The separation of chromatids is accomplished at the telophase which is followed by cytokinesis or the cell division.

1.4.2 PROPHASE

This is the longest part of mitosis, where centrioles migrate to the opposite ends of the cell and two poles are established leading to organization of cytoplasmic microtubules arranged parallel between the two poles. This organization is called spindle fibers along which chromosomes separate during karyokinesis. An interesting point to be mentioned here is that centrioles are not solely responsible for spindle fiber formation which occurs in plant cells too which lack these structures. During the migration of centriole, nuclear envelope disintegrates and disappears. The events occurring at the chromatin level during prophase are:

A. Condensation of chromatin which was in diffused state during interphase
    a. Condensation leading to visibility of chromosomes
    b. End of prophase results in adequate visibility of chromosomes to see that they are actually double-stranded structure joint at centromere.

An important point to note here is that the attached chromatin material are called sister chromatids which have identical genetic makeup.

1.4.3 METAPHASE

This phase got its name due to formation of the equatorial plane or metaphase plate perpendicular to the axis established by the spindle fiber. Chromosomes are arranged on the plane during this phase by binding centromere of each chromosome to spindle fiber by a structure called kinetochore. The chromosomes are in most condensed form at this stage thus, are greatly visible.

1.4.4 ANAPHASE

This is the shortest stage of mitosis, where sister chromatids of each chromosome separate from each other and migrate to other ends of the cell along the spindle fiber.

1.4.5 TELOPHASE

This is the final stage of mitosis which results in complete migration of sister chromatids to opposite poles followed by cytokinesis (division) and formation of two daughter cells. Since cells have to undergo another cycle of mitotic division the transition to interphase occurs during late telophase. The chromosome begins to uncoil and form diffuse chromatin and the nuclear envelope forms along with disappearance of spindle fiber. Mitosis is a continuous process with multiple regulatory mechanisms into place for efficient division. Uncontrolled division manifests itself into diseased condition. An example of such condition is cancer. Thus, the entire process is under the control of processes governed by genes present in the cell.

The various mechanisms and the molecules involved in the control of mitotic process have been dealt with in the subsequent part of this chapter.

1.5   GENETIC REGULATION OF CELL CYCLE

The cell cycle involves a series of carefully controlled events resulting in DNA duplication and cell division. This entire program of cell cycle has been conserved during evolutionary processes. This is evident by the fact that the all eukaryotic cell show similar mode of cell duplication. Progression through each of the four distinct phases (G1, S, G2, and M) is carefully controlled by the sequential formation, activation, and subsequent degradation or modification of a series of cyclins and their partners, the cyclin-dependent kinases (CDKs).[4] In addition, a further group of proteins, the cyclin-dependent kinase inhibitors (CDKIs) are important for coordination of each stage. The transition from one stage to the next is regulated at a number of checkpoints which prevent premature entry into the next phase of the cycle. The degradation of various cyclins occurs at each checkpoint and it is this mechanism together with interaction of the CDKIs which allows the cell to enter the next phase.

The role of a protein complex called DREAM has been explored in the G0, G1 phases of the cell cycle, especially the coordination of genes. [6,7] There are various complex mechanisms that influence the cascade of events involved in the transcriptional activation and deactivation associated with the cell cycle.[8]

1.5.1   CYCLINS

At least 13 cyclins have been identified in mammals; requirement of each one is at a different cell cycle stage. They all contain a homologous region of about 100 amino acids called the cyclin box and it is this region of the protein which binds to the appropriate CDK. Each cyclin undergoes a characteristic pattern of synthesis and degradation dependent on the stage of the cycle at which it acts. The cyclins are broadly classified into G1 and mitotic cyclins, according to the stage of the cycle during which they are produced. The G1 cyclins are relatively short-lived proteins and are degraded via their PEST sequences which lie C terminal to the cyclin box. The mitotic cyclins are longer lived and are degraded, prior to entry into mitosis, by proteinases via an ubiquitin-dependent pathway via the "destruction" box located N terminal to the cyclin box. Cyclin degradation results in CDK inactivation or tyrosine 15 by the wee1/mik1 protein kinase. At the end of G1, the product of the CDC25 gene activates the kinase by dephosphorylation of these residues.

CDKIs: There are at least seven different CDKIs in mammalian cells which belong to two different classes. The first class comprises p21, p27, and p57 which preferentially bind to the GI/S class of CDKs. The second class of CDKs, referred to as the INK4 (Inhibitor of CDK4) family, comprise an kyrin repeat proteins and include p15, p16, p18, and p19. These inhibitors act on cyclin D complexed either to CDK4 or CDK6.

CDKs: At least six mammalian CDKs have so far been identified. The CDKs are activated by forming complexes with cyclin partners and by a pattern of phosphorylation and dephosphorylation at specific residues. CDKs are activated by phosphorylation of a conserved threonine residue at position 160 and by cyclin binding. For example, Phosphorylation of the CDKs CDC2, CDK2, and CDK4 is carried out by the p40mol5 protein which in turn, is activated by cyclin H. In addition, phosphorylation also may cause inhibition of cyclin/CDK complexes, for example, the two CDKs, CDC2 and CDK2, are inactivated throughout interphase by phosphorylation on threonine at certain positions.

1.5.2   CONTROL OF THE CELL CYCLE

Each of the cyclin-CDK complexes, together with the CDKIs, are responsible for controlling different stages of the cell cycle by preventing progression through checkpoints in the presence of DNA damage.

1.5.2.1   LATE G1 CHECKPOINT

The D-type cyclins are linked to the regulation.

1.5.2.2   LATE G1 CHECKPOINT

The D-type cyclins are linked to the regulation of the first checkpoint at G1/S. They are synthesized in response to growth factors and are very short lived. The D-type cyclins are found in partnership with four kinases, CDK2, -4, -5, and -6 although CDK4 appears to be the main partner in most cell types. Activation of CDK4 by complexing with cyclin D and phosphorylation on threonine 160 drives cells through this checkpoint by phosphorylation of the Rb protein which releases the E2F transcription factor allowing it to activate genes necessary for DNA synthesis.

1.5.2.3   G1/S PHASE

The E-type cyclins are believed to act after the D type and to be important for the initiation of DNA replication. Cyclin E is expressed toward the end of G1, and complexes with CDK2 to activate it. As with cyclin D-CDK4, phosphorylation of the threonine residue (160) is necessary for activation. After cells have entered S phase, cyclin E is rapidly degraded and CDK2 is released to be complexed by cyclin A at the next stage. In natural circumstances, cells which have suffered DNA damage are prevented from entering S phase and are blocked at G1, a p53-dependent process through its it transcriptional regulation of the cyclin-dependent increased p21 level. It can then bind to number of cyclin-CDK complexes including cyclin D-CDK-4, cyclin E-CDK-2, and cyclin A-CDK-2 thereby preventing phosphorylation of pRb causing the cell cycle to arrest in G1.

1.5.2.4   S PHASE

Once cells enter S phase, a further set of cyclins and CDKs are required for continued DNA replication. In mammalian cells, cyclin A-CDK2 performs this function. Cyclin A is expressed from S phase through G2 and M. Cyclin A binds to two different CDKs. Initially during S phase, it is found complexed to CDK2 and during G2 and M, it is complexed to CDC2.

1.5.2.5   MITOSIS

Entry into the final phase of the cell cycle, mitosis, is signaled by the activation of the cyclin B–CDC2 complex. This complex accumulates during

S and G2 but is kept in the inactive state by phosphorylation of tyrosine 15 and threonine 14 residues, a process regulated by the weeI/mikI kinases. At the end of G2, the CDC25c phosphatase is stimulated to dephosphorylate these residues thereby activating CDC2. Cyclin B is located in the cytoplasm during interphase but is translocated to the nucleus at the beginning of mitosis. Cyclin B-CDC2 plays a major role in controlling the rearrangement of the microtubules in mitosis. In addition, the complex plays a role in disassembling the nucleus and allowing the cell to round up and divide.

There is one final checkpoint which occurs at the end of metaphase and at this point, the correct assembly of the mitotic apparatus and the alignment of chromosomes on the metaphase plate are monitored. Normal cells arrest at this point if there are any defects, whereas in tumor cells, abnormalities of spindle formation are found, suggesting that the checkpoint control is lost. Cyclin B is degraded as cells enter into anaphase. Reestablishment of interphase can then be initiated.

The various cyclin-dependent kinases, cyclins, and Cdk inhibitors (CKIs) have close interactions and cooperation with each other for optimum functioning of a cell. They have key roles in several processes such as metabolism, transcription, epigenetic regulation, stem cell self-renewal, neuronal functions, and spermatogenesis that have been reviewed by Lim and Kaldin, 2013.[9]

1.6 MEIOSIS

Meiosis also known as the reduction division is responsible for reducing the genetic material into half in gametes. Contrary to the production of diploid daughter cells in mitosis, haploid gametes or spores are produced in meiosis, which contain one member of each homologous pair of chromosomes. Meiosis is a very important phenomenon needed for continuation of genetic information from one generation to another. Broadly meiosis has been divided into two phases, meiosis I and meiosis II, which have been further subdivided into more subphases. The first division which occurs in meiosis I is referred to as reduction division since number of centromeres each representing one chromosome is reduces by one half. Meiosis II is an equational division because there, no change in the number of centromeres occurs.

1.6.1 MEIOSIS I

Meiosis I has been divided into prophase I, metaphase I, anaphase I, telophase I, and diakinesis (Fig.1.7).

Stages in meiosis

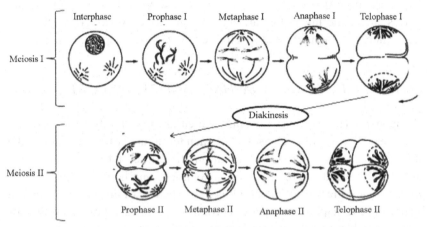

FIGURE 1.7 Cell undergoing meiosis: the first division is the reduction division, also called meiosis. The chromosome number is halved in this phase which has been divided into prophase I, metaphase I, anaphase I, and telophase I.

1.6.1.1 PROPHASE I

This phase is characterized by condensation of chromatin into visible chromosomes. This phase has been further subdivided into five stages, that is, leptonema, zygonema, pachynema, diplonema, and diakinesis. During leptotene stage, the chromatin material condenses though still in extended form and is visible. Shortening of chromosomes occurs in zygotene stage and alignment and pairing of homologous chromosomes occur. The homologous pairs formed are referred to as bivalents. Pachytene phase is characterized by further shortening of chromosomes and development of synaptic complex between two members of each bivalents. Thus, each bivalent contains four chromatids. This four-membered structure is called a tetrad. Each tetrad is composed of two pairs of sister chromatids and genetic recombination occurs between non-sister chromatids. In diplotene stage, each pair of sister chromatid in a tetrad begin to separate, but there are areas where chromatids are intertwined representing the point where genetic exchange has occurred between non-sister chromatids. Such area is called chiasma (or chiasmata). During diakinesis, the chromosomes pull apart but the non-sister chromatids remain associated at the chiasmata. As separation proceeds, the chiasmata move toward the end of the tetrad. In this stage, the nucleolus and nuclear envelope breaks down, the two centromeres of each tetrad attach to the newly formed spindle fibers.

1.6.1.2　METAPHASE I, ANAPHASE, AND TELOPHASE I

Metaphase I features chromosomes which are maximally shortened and thickened. The terminal chiasmata of each tetrad are still visible and is the only factor holding the non-sister chromatids. Each tetrad interacts with spindle fibres and aligns itself. At anaphase I, one half of each tetrad is pulled toward the pole of the dividing cell. The end of this phase is marked by presence of haploid number of chromosomes at each pole. During telophase I, nuclear membrane reappears and cell prepares for meiosis II.

1.6.2　MEIOSIS II

This phase is divided into prophase II, metaphase II, anaphase II, and telophase II. Prophase II shows the presence of one pair of sister chromatids attached by a common centromere. The dyads arrange themselves on the metaphase plate during metaphase II. Sister chromatids of each dyads are pulled apart to opposite poles during anaphase II. Telophase II is marked by presence of one member of each pair of homologous chromosomes at each pole. This is followed by cytokinesis and formation of four haploid gametes, with each gamete having mixture of genetic information of their ancestors.

　　Duro and Marston, 2015 have reviewed various factors involved in the segregation of the homologous chromosomes.[9] These molecular mechanisms involved in the segregation are of utmost importance in the maintenance of ploidy levels and segregation of chromosomes.

1.7   SUMMARY

After dealing with the cell cycle processes, it becomes pertinent to further dig into details of other molecular processes occurring within the basic unit of life, that is, the cell. All this starts with understanding the language in which a cell is guided through all these events. The language is the genetic cone and its alphabets are the various nitrogenous bases which would be dealt with in the subsequent part of this book.

1.8   REVIEW QUESTIONS

1) Describe the packing of chromatin.
2) Explain the various aspects of the cell cycle.
3) Highlight the similarities and differences between a prokaryotic and a eukaryotic cell.
4) Elucidate the properties of virus that places it at the threshold of life and death.
5) What is the basis of the statement "mitochondria and chloroplast are symbiotic prokaryotes"?
6) Compare and contrast between eubacteria and archaebacteria.

KEYWORDS

- **cell**
- **archaebacteria**
- **eukaryotes**
- **animal cell**
- **cell cycle**
- **bacteria**
- **virus**
- **mitosis**
- **meiosis**

REFERENCES

1. Karp G. *Cell and Molecular Biology Concepts and Experiments.*
2. Woolverton, C. J.; Willey, J.; Sherwood, L. *Prescott's Microbiology.*
3. Stanier, R. Y.; John L. Ingraham, J. L.; Wheelis, M. L. Painter, P. R. *General Mirobiology*, 5th ed.
4. Klug, W. S.; Cummings, M. R. *Essentials of Genetics*, 5th ed.
5. Deng, X., et al. Cytology of DNA Replication Reveals Dynamic Plasticity of Large-scale Chromatin Fibers. *Current Biol.* **2016**. DOI:10.1016/j.cub.2016.07.020.
6. Sadasivam, S.; DeCaprio, J. A. The DREAM Complex: Master Coordinator of Cell Cycle-dependent Gene Expression. *Nature Rev. Cancer* **2013**, *3*, 585–595.
7. Bertoli, C.; Skotheim, J. M.; de Bruin, R. A. M. Control of Cell Cycle Transcription During G1 and S Phases. *Nat. Rev. Mol. Cell Biol.* **2013**, *14*, 518–528.
8. Lim S., Kaldis, P. Cdks, Cyclins and CKIs: Roles Beyond Cell Cycle Regulation. *Development* **2013**, *140*, 3079–3093. DOI: 10.1242/dev.091744.
9. Eris Duro E.; Marston, A. L. From Equator to Pole: Splitting Chromosomes in Mitosis and Meiosis. *Genes Dev.* 2015, *29*, 109–122.

GENES AND GENETIC CODE

CONTENTS

ABSTRACT

Molecular basis of heredity has long been a subject of discussion. The central dogma theory forms the pivot for explanation of processes such as transcription and translation. DNA/RNA forms the genetic basis of heredity. Nucleotides are the alphabets of this genetic language. Any change in the sequence of this may lead to changes in the cellular characteristics which is referred to as mutation. Though mutation is a phenomenon occurring regularly but is taken care of by various cellular mechanisms. Genetic code its ambiguity, replication, transcription, translation, and mutation has been dealt with in this chapter.

2.1 THE CENTRAL DOGMA

2.1.1 THE CENTRAL DOGMA OF LIFE

In each and every living being all the life process occurs at all time at molecular level. The study of these molecular processes has given rise to a new branch of science that is molecular biology. This study has provided many insights into the life processes which we were oblivious of. All these developments have intrigued human mind to a great deal and has opened new avenues for research. All this started with the deduction of structure of deoxyribonucleic acid (DNA) by Watson and Crick. This giant leap led to deciphering of genetic code and subsequently the processes of replication, transcription, and translation was elucidated. This chapter deals with the aspects of this new branch of biological science. As the genetic coder dictates the mode of conduct of all the processes, any change in its constitution may cause undesirable modification in the phenotype. This aspect of molecular biology has also been dealt with in this chapter with reference to mutation.

DNA contains the complete genetic information that defines the structure and function of an organism. Proteins are formed using the genetic code of the DNA. Three different processes are responsible for the inheritance of genetic information and for its conversion from one form to another (Fig. 2.1) are replication, transcription, and translation which have been defined and described below:

1. *Replication:* a double-stranded nucleic acid is duplicated to give identical copies. This process perpetuates the genetic information.
2. *Transcription:* a DNA segment that constitutes a gene is read and transcribed into a single-stranded sequence of ribonucleic acid (RNA). The RNA moves from the nucleus into the cytoplasm.
3. *Translation:* the RNA sequence is translated into a sequence of amino acids as the protein is formed. During translation, the ribosome reads three bases (a codon) at a time from the RNA and translates them into one amino acid.

In eukaryotic cells, the second step (transcription) is necessary because the genetic material in the nucleus is physically separated from the site of protein synthesis in the cytoplasm in the cell. Therefore, it is not possible to translate DNA directly into protein, but an intermediary must be made to carry the information from one compartment to another.

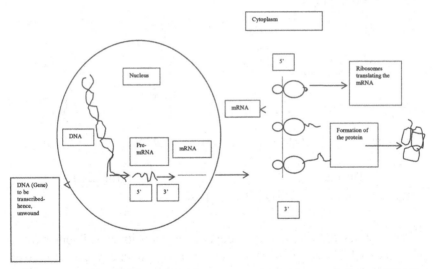

FIGURE 2.1   Flow of information from gene to protein, that is, central dogma of life. The genetic information present in the DNA which is present in the nucleus is transcribed into mRNA. mRNA comes out of the nucleus to act as a template for translation that is protein synthesis by the ribosomes.

In certain acellular organisms such as viral RNA also acts as genetic material, which may or may not serve as a template for protein synthesis.

If protein can be directly translated from the genetic material then the RNA is referred to as +(plus) strand otherwise –(minus) strand which acts as a template for synthesis of + strand RNA

2.1.2 GENOME AND STRUCTURE OF NUCLEIC ACID

The three pillars of molecular genetics are DNA, RNA, and protein. As already discussed that DNA serves as genetic material in all organism except for some virus where RNA serves as a genetic material. In a cell the DNA, RNA, and protein work in proper coordination to achieve the various function assigned to these molecules. DNA exists in nucleus as a double-helical structure. Each strand of the double helix is a polymer made up of nucleotides.

The Nucleotides of DNA

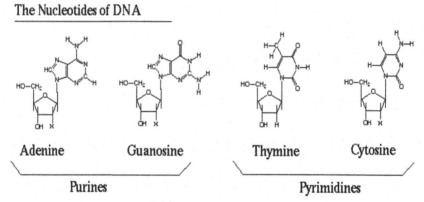

FIGURE 2.2   Structure of purines and pyrimidines.

Nucleotides are of four types depending upon its constituent nitrogenous base which are adenine (A), guanine (G), cytosine (C), and thymine (T). In RNA uracil (U) is present instead of thymine. A and G are called purines, C and T are called pyrimidines (Fig. 2.2). The two strands of the double helix are complimentary to each other where a double bond is formed between A and T and a triple bond is formed between G and C base pairs (refer to Fig. 2.3). The type of bonds formed is hydrogen bond which is very important for genetic function.

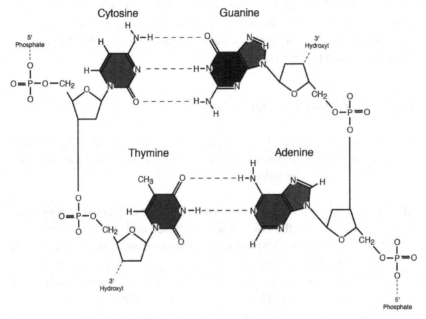

FIGURE 2.3 Hydrogen bond formed between guanine, cytosine, and adenine, thymine present on the complimentary strands of the double helix of DNA.

2.1.3 HISTORY AND ORIGIN OF GENETIC CODE

History of mankind had a huge impact of genetics in its evolution. Genetics as a discipline is a recent development, but its impact can be echoed throughout the evolutionary upgradation of humans with respect to their surroundings as can be seen in breeding of crops and animals. Cultivation of plants such as wheat, rice, and maize which are major dietary components throughout the world began along with the early settlements which followed breeding and rearing of domestic animals such as oxen, horses, camels etc. this was made possible by careful observation of desirable traits its manipulation and its propagation in prehistoric times. Genetics is the only thread which unites various disciplines such as cell biology, molecular biology, physiology, biochemistry, evolutionary biology, ecology etc.

The genetic code is a linear sequence of DNA which controls and regulates the end product of the gene, that is, the protein. In some cases, RNA can also act as genetic code as seen in case of many viruses. The message encoded by DNA is deciphered in the system in the form of RNA also called messenger RNA by transcription followed by protein synthesis known as translation. To translate a particular messenger (m) RNA, it has to get associated with ribosomes. This decoding of DNA to RNA to protein is known as central dogma of life.

2.1.4   THE GENETIC CODE

It is important to understand the genetic code prior to unravel the mystery of transcription and translation process. The various unique characteristics of genetic code are listed below:

1. The genetic code is always written in a linear form. The alphabets of the code are the ribonucleotide bases and the letter formed is "m RNA" whose sequence is complimentary to the nucleotide bases of DNA.
2. The three-nucleotide sequence is referred to as codon, which codes for an amino acid. This codon is called triplet codon. The triplet codon is specific for each amino acid (unambiguous).
3. Degeneracy of codon: few amino acids can be specified by more than one triplet codon.
4. The code also consists of a start and a stop codon which defines the beginning and end point of the translation process.
5. No punctuation in the form of comma or full stop is incorporated while writing the code.
6. The code is nonoverlapping.
7. The code is universally used by all living organisms with very little exception where modified bases are found in certain bacteria and virus.
8. Besides the universal meaning of the genetic code, in certain organelles in various organism, their meaning tends to differ as illustrated in Table 2.1.

The genetic code of *Escherichia coli* was completed in 1968 and later determined for many other organisms, which was found to be same giving an impression that genetic code is universal. Then further studies led to discovery of exception to universal genetic code as listed in Table 2.1, which shows the preference of certain codons over others during usage. This phenomenon was named as codon usage bias.

TABLE 2.1 Exceptions to the Universal Genetic Code.

Organism	Normal codon	Usual meaning	New meaning
Mammalian	AGA, AGG	Arginine	Stop codon
Mitochondria	AUA	Isoleucine	Methionine
	UGA	Stop codon	Tryptophan
Drosophila	AGA, AGG	Arginine	Serine
Mitochondria	AUA	Isoleucine	Methionine
	UGA	Stop codon	Tryptophan
Yeast	AUA		
Mitochondria	UGA	Isoleucine	Methionine
	CUA, CUC	Stop codon	Tryptophan
	CUG, CUU	Leucine	Threonine
Higher plant	UGA	Stop codon	Tryptophan
Mitochondria			
Protozoan nuclei	UAA, UAG	Stop codon	Glutamine
Mycoplasma	UGA	Stop codon	Tryptophan

2.1.5 NATURE OF GENETIC CODE

As it has been already discussed about the availability of extra words in codon language, there must be some purpose attached to its existence. Some codon do not code for any amino acid, they are called nonsense codons and used by the system for termination of translation, these codon are UAA (ochre), UGA (opal), and UAG (amber), the stop codons were studied in mutants of bacteriophage T4 and λ (lambda) which infect the bacteria *E. coli*. Among these stop codons "amber" was first discovered. Viruses having amber mutation were able to infect strains of bacteria carrying amber suppressor mutation. This suppressor mutation allowed the recovery of lost function in virus. Similarly, opal and ochre codons were discovered in bacteria harboring suppressor mutation for these two codons. Amber mutations were the first nonsense mutations to be discovered by Richard Epstein and Charles Steinberg.

To address the question posed by the availability of extra codons, "degeneracy" of the codons was proposed. Degeneracy of codons means existence of multiple codes for an amino acid. Some amino acids such as methionine and tryptophan have a single codon whereas multiple codons exist for other amino acids (Table 2.1) but the point to be kept in mind is that for any specific codon, only a single amino acid is indicated though there are exception to this rule too, which is referred to as *unambiguous* genetic code.

It is a general understanding that the third nucleotide in the codon is less important than the first two in determining the incorporation of specific amino acid in the growing peptide chain this phenomenon is known as "Wobble hypothesis." This hypothesis was proposed by Crick in 1966 postulating that the first two ribonucleotides of the triplet codon are more critical than the third in attracting tRNA. This wobble or flexibility provides for the degeneracy of codons were a tRNA can recognize multiple codons. This also explains the availability of codons in much excess than needed and one amino acid can be coded by multiple codons.

The other very important property of genetic code is that it is nonoverlapping in nature, that is, there is no comma or full stop between the codon and the message is read in a continuous fashion till a nonsense codon appears. With a few exceptions, the genetic code is considered to be universal.

2.1.6 CODON USAGE BIAS

A very important characteristic of genetic codon is its degeneracy which has a meaning that multiple codons specify the same amino acid that is they code for the same amino acid (Table 2.2) such codons are referred to as synonymous codon. It was assumed that synonymous codons for any amino acid would appear randomly distributed along the genes. But this is not so, it has been observed [1-3] that some codons are repeatedly preferred over others. Thus, codon usage bias refers to the phenomenon where specific codons are used more often than other synonymous codons during translation of genes. Synonymous codons do not appear to be used equally and randomly to code for an amino acid. Some codons are repeatedly preferred over others; this phenomenon is termed codon usage bias. Codon usage frequencies, in fact, vary among genomes, among genes, and within genes.

TABLE 2.2   The Genetic Code as in Messenger RNA.

First nucleotide	Second nucleotide				Third nucleotide
	U	**C**	**A**	**G**	
U	Phe	Ser	Tyr	Cys	U
	Phe	Ser	Tyr	Cys	C
	Leu	Ser	Stop codon	Stop codon	A
	Leu	Ser	Stop codon	Trp	G
C	Leu	Pro	His	Arg	U
	Leu	Pro	His	Arg	C
	Leu	Pro	Gin	Arg	A
	Leu	Pro	Gin	Arg	G
A	Ile	Thr	Asn	Ser	U
	Ile	Thr	Asn	Ser	C
	Ile	Thr	Lys	Arg	A
	Met Start codon	Thr	Lys	Arg	G
G	Val	Ala	Asp	Gly	U
	Val	Ala	Asp	Gly	C
	Val	Ala	Glu	Gly	A
	Val	Ala	Glu	Gly	G

Synonymous codons with different preferences are used in different organisms (Table 2.1). It has been observed that the most common codon corresponds to the highest translational efficiency in heterologous expression. Disadvantages of this method are:

a)   If only one codon is used to encode each amino acid, there is only a single possible DNA sequence with which to encode a specific protein.

b)   It also eliminates any flexibility in other design criteria such as the elimination or incorporation of restrictions sites, repetitive elements within the sequence which can compromise stability or sequences that could form structures at or around the site of translational initiation.

c)   Overuse of some codons may also result in significant amino acid mis-incorporation as suggested by Kurland and Gallant, 1996, which might lead to compromising the function of the protein.

d)   However, the codon usage may not be optimal for expression and there are several evidences suggesting that genes designed using common codons are not correlated with high protein expression.

Till date in all cases studied so far, expression is highly correlated with codon usage but does not show a general preference for use of codons used at highest frequency in the genome or in the highly-expressed gene subset of the host. Much further research is needed to fully understand the nature of these effects bias.

2.1.7 THE TRIPLE NATURE OF GENETIC CODE

Discovery of mRNA led to elucidation of decimation of information stored in DNA to mRNA to protein synthesis. But the question faced was that how this information in the form of four letters (the four nucleotides) specified the 20 words for amino acids. The triplet nature was derived from experiments done on T4 bacteriophage by Francis Crick and his group. They studied frameshift mutation (result of deletion or addition of a nucleotide) within a gene and the mRNA transcribed by it. The gain or loss of a nucleotide results in shift in reading frame during translation. It was observed that loss or gain of one or two nucleotide cause mutation but when this event of deletion or addition involved three nucleotides the reading frame was reestablished. This experiment firmly established the fact that the codes are triplet in nature. Moreover, if the codes were represented by two codons then the mathematically four codons would code for only 16 amino acids ($4^2 = 16$), while if the codes were in triplet then four codons would code for more than 20 words which were needed ($4^3 = 256$). This was argued by Sidney Brenner in early 1960s.

2.1.7.1   DECODING THE GENETIC CODE: NIRENBERG AND KHORANA

Marshall Nirenberg and J. Heinrich Matthaei were the pioneers in deciphering the genetic code. They deciphered the first genetic codon in 1961 by busing an *in vitro* cell-free protein synthesis system using an enzyme polynucleotide phosphorylase. This system was used for the production of synthetic mRNAs which are templates for polypeptide synthesis in cell-free system. The *in vitro* mixture contained essential factors for protein synthesis, that is, ribosomes, tRNA amino acid, and other molecules necessary for protein synthesis.

An enzyme needed for RNA template synthesis polynucleotide phosphorylase was isolated from bacteria in 1955 by Marianne Grunberg-Manago and Severo Ochoa. This enzyme catalyzes the reaction of RNA synthesis (Fig. 2.4).

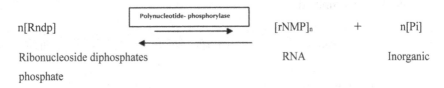

$n[Rndp]$ ⟶ Polynucleotide- phosphorylase ⟶ ⟵ $[rNMP]_n$ + $n[Pi]$

Ribonucleoside diphosphates   RNA   Inorganic
phosphate

FIGURE 2.4   *In vitro* synthesis of RNA.

In vitro, this reaction can be forced in the direction of RNA synthesis. Polynucleotide phosphorylase does not need a DNA template unlike RNA polymerase, thus addition of nucleotide becomes a random process dependent on the relative concentration of nucleotides (ribonucleoside diphosphates) in the reaction mixture. Thus, availability of cell-free system for protein synthesis and synthetic mRNA gave the much needed momentum to decipher the genetic code.

Nirenberg and Matthaei synthesized RNA homopolymers with single type of nucleotide, so the mRNA synthesized by them was either UUUUUUU....., AAAAAAAAA......, CCCCCCCCC...., or GGGGGGG...... Each mRNA was tested to determine which amino acid they coded for and it was concluded that poly U (poly uridylic acid) directed the incorporation of only phenylalanine. Thus, they were able to determine the triplet codon for phenylalanine which was UUU. Similarly, they found that AAA codes for lysine and CCC codes for proline, poly G was not adequate template. Thus, this work proved to be the foundation on which the entire structure of genetic code was built. This work was replicated using RNA heteropolymer by Nirenberg and Matthaei and Ochoa. In this two or more ribonucleoside diphosphate were added to form the artificial message keeping in mind if the relative proportion of each type of ribonucleoside diphosphate was known then frequency of any particular triplet codon occurring in synthetic RNA could be predicted. This synthetic mRNA if subjected to translation in cell-free system and the percentage of any amino acid present in the translated protein then this could be used to predict the composition of triplets specifying particular amino acids. In 1960, a scientist of Indian origin Hargovind Khorana was able to chemically synthesize long RNA molecule consisting of short sequences repeated several times. He first created very short sequences of di-, tri-, and tetranucleotides which were replicated several times and then joined enzymatically to form long polynucleotides (Fig. 2.5). This synthetic message when added to cell-free system predicted number of amino acid incorporated was upheld.

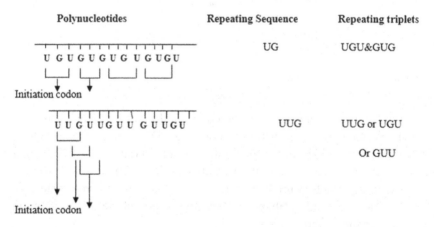

FIGURE 2.5   Experiment of Hargovind Khorana to chemically synthesize RNA molecule by conversion of di- and trinucleotide into repeating copolymers enzymatically.

2.1.7.2   START AND STOP CODONS

Initiation of translation occurs by insertion of initiating amino acid in the polypeptide chain which is methionine except in bacteria where it is

N-formylmethionine which is coded by triplet codon AUG. This codon is known as the initiation codon which is very specific to the translation process. An initiated process must come to an end, and the end of protein synthesis is termination codon which is UAG, UAA, and UGA which do not code for any amino acid. These codons are not recognized by tRNA molecule and are also known as stop codons (Table 2.2).

2.2   RELATIONSHIP BETWEEN GENES AND PROTEINS

Insight into this relationship was made by studying the symptoms of rare disease. Archibald Garrod made a significant contribution by studying inherited disease alkaptonuria, conspicuous by darkening of their urine when it is exposed to air. This phenomenon occurred because of lack of an enzyme in their blood that oxidized homogentisic acid, a compound formed during the breakdown of the amino acid phenylalanine and tyrosine. Homogentisic acid accumulates in the excreted urine and darkens in color when oxidized by air. Thus, Archibald Garrod had discovered relationship between specific gene, enzyme, and metabolic condition which he called "inborn error of metabolism." This study was further carried on by George Beadle and Edward Tatum. They used *Neurospora* a tropical bread mold as experimental model, which has a capability to grow in a simple medium containing organic carbon (sugar), inorganic salts, and biotin (vitamin B). Since its growth requirement was very simple, it must be having the capability to synthesize all the required metabolites. So, Beadle and Tatum irradiated mold spores and distributed them so that they produce genetically identical population. This was done keeping in mind the organism with such wide spectrum of synthetic ability should be sensitive to enzyme deficiencies, which could be detected using experimental protocols. By this experiment, they were able to isolate cells unable to grow on minimal media. Two such isolates were identified one needed pyridoxine (vitamin B6) and the other needed thymine (vitamin B1). These isolates were mutant for the gene which resulted in an enzyme deficiency preventing the cells from catalyzing a particular biological reaction. Thus, this experiment led to the hypothesis of "one gene-one enzyme." Later on, it was found that an enzyme may contain several polypeptide chains encoded by different gene, then the hypothesis was modified to "one gene-one polypeptide chain." The question about the molecular nature of defect in a protein due to genetic mutation from findings of Beadle and Tatum was responded to by Vernon Ingram (1956) by his work on sickle cell anemia which was a result of mutation in β-chain of hemoglobin. This mutation caused the substitution of a valine in the sickle cell hemoglobin for a glutamic acid. This work clearly demonstrated that a mutation in a single gene had caused this substitution in amino acid sequence of a single protein.

> *Prototroph: microorganism capable of fulfilling their nutritional and growth requirement in minimal media.*
>
> *Auxotroph: microorganism unable to support their growth on minimal media and having specific nutritional requirements.*

The main three life processes are replication, transcription, and translation which form the three pillars of the theory of "central dogma of life." All the three processes are interdependent. Replication is required to decimate the information to daughter cells, which need the participation of various proteins formed as a result of transcription and translation process. Understanding of all these things has been facilitated by working of great minds some of whom contribution has been mentioned in this chapter.

2.3 REPLICATION

Replication is the process of duplication of genetic material. This is a unique feature of all living beings for their propagation which nullifies its chances of loss into oblivion. RNA is considered to be the first material used as carrier of genetic material. Later during the course of evolution, DNA was selected as the preferred carrier of genetic material in most of the organism. This also led to increase in the complexity of replication process which required a number of auxiliary molecules. To replicate there is an absolute requirement of the cell which provides adequate environment for effective execution of replication process.

Elucidation of DNA structure by Watson and Crick in 1953 was accompanied with a proposal for self-duplication of DNA. The double strands of DNA are held together by hydrogen bonds between the base pairs, two bonds between adenine and thymine, three bonds between guanine and cytosine (Fig. 2.6).

Chemical Structure of DNA

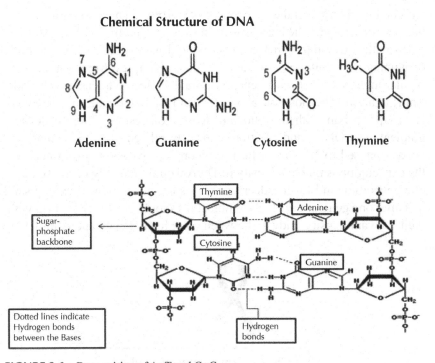

FIGURE 2.6   Base pairing of A=T and G≡C.

2.3.1 SEMICONSERVATIVE MODE OF DNA REPLICATION

Initially, the mode of DNA replication was not known and various theories were put forward, they were conservative, semiconservative, and disruptive. In each of daughter nuclei, the DNA duplex formed is made up of one strand from the parental origin and the second is completely newly synthesized strand; it is known as semiconservative replication (Fig. 2.7). In conservative mode, the two original strand would remain together in one daughter cell and the newly synthesized double strand DNA would be in another daughter cell. The disruptive mode would be characterized by presence of strand composed of mixture of parental strand and newly synthesized strand that is each strand is a composite of old and new DNA.

Mode of replication was first studied by Meselson and Stahl in the year 1958 at California Institute of Technology. The study was done in bacteria. The bacteria were grown for many generations in media containing N^{15}

ammonium chloride as the sole source of nitrogen. Thus, the nitrogenous bases contained only the heavy nitrogen isotope. Cultures of these bacteria laden with heavy nitrogen were washed and incubated in fresh medium containing N^{14} containing nitrogen source. Samples were obtained at regular intervals for several generations and subjected to cesium chloride centrifugation. The position at which band of DNA molecule is found is in equilibrium with the buoyant density of cesium chloride density gradient. The DNA band position is directly related to the percentage of N^{15} as opposed to N^{14} atoms. In a semiconservative mode of replication, the expectation is that the density of DNA would gradually decrease up to one generation and further decrease in second generation. It can be said that after one generation all DNA molecules should be N^{15}–N^{14} hybrid with their buoyant density midway between completely heavy and completely

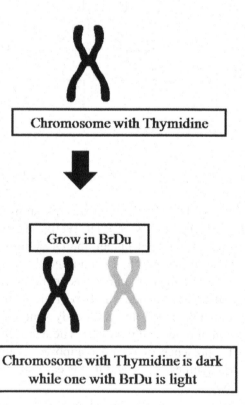

FIGURE 2.7 Various proposed mode of DNA replication (a) conservative (b) semi-conservative, and (c) disruptive.

light DNA. After subsequent generations, the DNA only contains the light isotope as the time of growth increases in the medium containing light isotope of nitrogen, the abundance of light DNA molecule will increase. If the replication is semiconservative the original heavy parental strand should remain intact but occupy smaller percentage with subsequent generations (Figs. 2.7 and 2.8).

Later the semiconservative replication was also demonstrated in eukaryotes in which the mammalian cells were cultured and allowed to go through two rounds of replication in bromodeoxyuridine (BrdU) which is a base analog incorporated into DNA in place of thymidine. Thus, incorporation of BrdU in chromatids of first and second-generation DNA verifies the semiconservative mode of replication (Fig. 2.7). Likewise, autoradiography analysis which is a cytological technique to see the location of a radioactive isotope in a cell established the semiconservative mode of replication. Work done by J. Herbert Taylor, Philips Woods, and Walter Hughes experimented on root tips of *Vicia faba* (broad bean) using ^{3}H-thymidine (radioactive precursor of DNA) in culture medium monitored the replication process. They placed a photographic emulsion over a section of root tip and stored this in a dark place. The slide was developed as a photographic film is developed. Presence of dark spots on the film located the newly synthesized DNA. Root tips were grown from one generation in ^{3}H-thymidine and then transferred to the nonradioactive

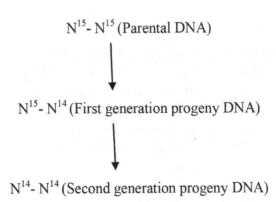

$$N^{15}\text{-}N^{15}\ (\text{Parental DNA})$$

$$N^{15}\text{-}N^{14}\ (\text{First generation progeny DNA})$$

$$N^{14}\text{-}N^{14}\ (\text{Second generation progeny DNA})$$

FIGURE 2.8 Semiconservative replication as verified using radioactive nitrogen. Parental cells previously incorporated with heavy isotope of nitrogen was shifted to medium containing normal nitrogen. Cell division led to incorporation of lighter isotopes of nitrogen into progeny thereby confirming semiconservative mode of replication.

medium, where cell division continued. At the end of each generation cells were arrested in metaphase using colchicine and chromosomes were observed by autoradiography. The result confirmed the semiconservative mode of replication (Fig. 2.8).

Study of replication in bacterial cells made a great progress due to various factors:

Availability of mutants lacking the ability to synthesize a particular protein is essential for replication but the cache is that such mutations are manifested at an elevated temperature, which is called nonpermissive temperature. When the cells are grown at lower temperature also called permissive temperature, the mutant protein carries its required activity and cells continue to grow and divide.

Availability of *in vitro* systems to study the process of replication using purified cellular components.

All these studies have revealed the presence of plethora of proteins needed for replication as elaborated in Table 2.3.

TABLE 2.3   Proteins and Genes of *E. coli* DNA Replication.

Protein	Gene	Function
DnaA	dnaA	Initiation of chromosome division binds to origin of replication
Helicase	dnaB	Unwinds the double helix
DnaC	dnaC	Loading of DNA helicase
SSB	ssb	Single strand binding protein
Primase	dnaG	Synthesis of RNA primers
RNase H	rnhA	Partial removal of RNA primer
Pol I	polA	Polymerase I fill gaps between Okazaki fragments
Pol III		DNA polymerase III holoenzyme
α	dnaE	Strand elongation
ϵ	dnaQ	Kinetic proofreading
Θ	holE	Unknown, part of core enzyme
β	dnaN	Sliding clamp
τ	dnaX	Dimerization of core enzyme
Υ	dnaX	Loading of sliding clamp
δ	holA	Loading of sliding clamp
$\delta`$	holB	Loading of sliding clamp
χ	holC	Loading of sliding clamp

TABLE 2.3 *(Continued)*

Protein	Gene	Function
ψ	holD	Loading of sliding clamp
DNA Ligase	lig	Seals nicks in lagging strand
DNA Gyrase		Introduces negative supercoils
α	gyrA	Makes and seals double-strand breaks in DNA
		ATP using subunit
β	gyrB	
Topoisomerase IV		Makes and seals double-strand breaks in DNA
A	parC	ATP using subunit
B	parE	

There are many more proteins observed in case of eukaryotic cell which play various important role in replication, thus apart from the aforesaid mentioned proteins other proteins involved in replication are present in other organisms, for example, telomerase in eukaryotes (refer to Table 2.4). It has also been observed that there are few differences in the properties of certain proteins in a prokaryotic and a eukaryotic cell as mentioned in Table 2.4. Though the process of replication is central to all the living cells but the path through which it is accomplished is the property of that cell, that is, whether it is a prokaryote or a eukaryote.

TABLE 2.4 A Comparison of Replication Process in Prokaryote and Eukaryote.

Steps	Prokaryotic cell	Eukaryotic cell
DNA double helix unwinding at origin (s) of replication	Helicase (ATP required)	Helicase (ATP required)
Stabilization of unbound template strands	Single strand binding proteins (SSB)	Single strand binding proteins (SSB)
Synthesis of DNA		
Leading strand	DNA polymerase III	DNA polymerase delta
Lagging strand (Okazaki fragment)	DNA polymerase III	DNA polymerase alpha
Removal of RNA primers	DNA polymerase I	Unknown
Replacement of RNA with DNA	DNA polymerase I	Unknown

TABLE 2.4 *(Continued)*

Steps	Prokaryotic cell	Eukaryotic cell
Joining of Okazaki fragments	DNA ligase I	DNA ligase I
Removal of supercoils preceding replication fork	Topoisomerase II (DNA gyrase)	Topoisomerase II
Synthesis of telomers	Not required	Telomerase

2.3.2 INITIATION OF REPLICATION

Replication starts with the separation of double-stranded DNA helix and synthesis of a pair of complimentary strands. Autoradiography developed by John Cairns in early 1960s proved to be very vital to study the replication phenomenon. This technique was done with gently lysed bacteria spread on a surface for the study. The bacterial cells were grown on a medium containing³H-thymidine before chromosome preparation. These prepared cells were observed under light microscope to see the outline of the labeled DNA strand. Such observation confirmed the circular nature of the bacterial chromosome along with the overall pattern by which replication occurs. The pattern formed was termed as theta (Θ). Each theta structure is composed of three distinct lengths of DNA (Fig. 2.9). These are formed of un-replicated chromosomal part and a pair of daughter molecules in the process of formation. The replicated segment is jointed to the nonreplicated segment at the replication fork where the double helix is undergoing separation and incorporation of nucleotides into the newly synthesized strand is occurring.

Replication begins at a specific site on chromosome called *origin*. The origin of replication in *E. coli* is called *oriC* having 245 base pairs to which multiple proteins bind to initiate the process. The replication fork appears at the origin of replication and moves with progressing replication. The length of DNA which is replicated after initiation at single origin is a unit called replicon. Replication can be unidirectional or bidirectional (refer to Fig. 2.9). If it is bidirectional two replication forks will appear migrating in opposite direction away from the origin. There is only one origin of replication in bacteria and bacteriophages that is they have one replicon. Autoradiography results have demonstrated the bidirectional mode of replication starting at *oriC* and terminating at *ter* (termination region).

Replication is accomplished by an enzyme known as DNA polymerase. This enzyme occurs in multiple forms in a cell with distinct functions.

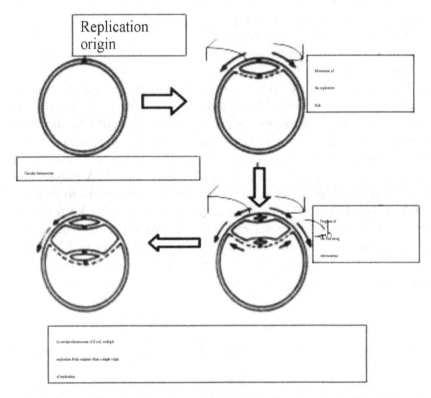

FIGURE 2.9   Bidirectional semiconservative mode of replication in *E. coli* showing the origin of replication, movement, and progress of fork as replication proceeds. The formation of a theta structure is a characteristic of bacterial replication.

2.3.3 DNA POLYMERASE

Bacteria have three DNA polymerases I, II, and III having specific functions. The basic function of all the polymerases is to add deoxyribonucleotides to the growing end of single-stranded RNA primer as well as exonuclease activity which is removal of terminal nucleotides one at a time. In *E. coli*, DNA polymerases III plays primary role in replication and is also called replicase. This enzyme is also referred to as holoenzyme because the active protein consists of a number of accessory subunits in addition to the polymerase itself. The catalytic part consists of three subunits (α, ε, and Θ) and noncatalytic β subunit which is responsible for keeping the polymerase associated with the DNA template, thus it is also referred to as β clamp (Fig. 2.10). β clamp also allows the enzyme to move from one nucleotide to other without leaving the template.

DNA polymerase II also has polymerase activity and no loss of function has been observed in mutants for this. DNA polymerase I is the enzyme which removes RNA primers at the 5′end of Okazaki fragment and replaces them with DNA. Apart from this polymerase I also has 3′ 5′ and 5′ 3′ exonuclease activity. Most nucleases are specific for either DNA or RNA, 5′ 3′ exonuclease activity of DNA polymerase I can degrade either DNA or RNA, giving it the property to degrade RNA primer. In Okazaki fragments, apart from removing RNA primer are also involved in filling the gaps between two fragments with appropriate nucleotides and the end sealed with ligase (refer to Fig. 2.10) there is a lot of difference between a prokaryotic and eukaryotic DNA polymerase (Table 2.5).

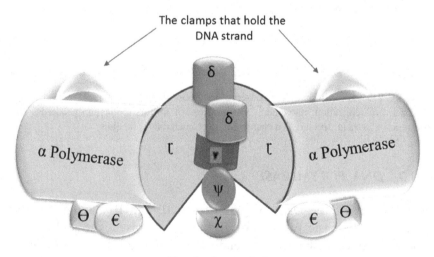

FIGURE 2.10   DNA polymerase I.

Replication starts with the separation of DNA strands. This separation leads to formation of the replication fork. Since DNA is a helical structure unwinding and separation of strands is a complex process involving a number of proteins. DNA gyrase plays a very important role in changing the positively supercoiled DNA into negatively supercoiled DNA by cleaving both the strands of DNA duplex and passing a segment of DNA through the double-stranded break to the other side and then sealing the cut. After this DNA helicase and single strand binding protein (SSB) come into play.

TABLE 2.5 Eukaryotic and Prokaryotic DNA Polymerases at a Glance.

E. coli	I	II	III		
Polymerization: $5' \rightarrow 3'$	+	+	+		
Exonuclease activity:					
$3' \rightarrow 5'$	+	+	+		
$5' \rightarrow 3'$	+	−	−		
Mammalian cells*	α	β[†]	Γ	δ	ε
Polymerization: $5' \rightarrow 3'$	+	+	+	+	+
Exonuclease proofreading activity:[‡]$3 \rightarrow 5'$	−	−	+	+	+
Cell location:					
Nuclei	+	+	−	+	+
Mitochondria	−	−	+	−	−

*Yeast DNA polymerase I, II, and III are equivalent to polymerase α, β, and δ, respectively.

Helicases are involved in unwinding of the duplex by breaking the hydrogen bonds which holds the two strand together. Helicase in *E. coli* is a product of *dnaB* gene. The helicases firstly attach to the DNA at its origin of replication and moves in $5' \rightarrow 3'$ direction along the lagging strand template (Fig. 2.11) unwinding the DNA helix. SSBs facilitate the separation of strands by attaching to the single-stranded DNA and keeping it in extended state thus preventing its rewinding. Synthesis of Okazaki fragment is initiated by an enzyme called primase which synthesizes short RNA primers periodically at specific sites along the lagging DNA strand. The primase and helicase are closely associated with bacteria forming primosome. Initiation, elongation, and completion of Okazaki fragments are a very rapid process which involves DNA polymerase III. Replication initiates at ori C and commences at *ter* where the two replication fork meets, the replicative enzymes disengage from DNA association and the two daughter chromosomes separate. This process is mediated by topoisomerase. The process of replication is controlled at the origin depending on the binding of regulatory protein, the machinery once started continues till entire chromosome has replicated. Survival of an organism depends upon the accuracy of replication process; the mistake in duplication manifests itself in transcription and translation which might prove to be lethal for the cell. The rate of mistake while copying the genetic material in *E.coli* has been estimated to be 10^{-9}. Incorporation of nucleotide in the growing chain is the function of DNA polymerase, which sometimes makes mistake in this by selecting incorrect nucleotide. Still, the process of DNA replication has great fidelity due to exonuclease activity of polymerases, which checks and excises the incorrect base from the growing strand, which is called *mismatch repair* (Fig. 2.12).

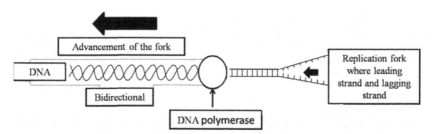

FIGURE 2.11   DNA Replication proceeding in forward and reverse direction forming leading and lagging strand (Okazaki fragments).

Replication in a eukaryotic cell is very similar to that of a prokaryotic cell with requirement of unwinding proteins, single-stranded DNA binding proteins, topoisomerases, primase, DNA polymerase, and ligase. In eukaryotic cell, five types of DNA polymerases have been identified. The list and function of the polymerases are listed in the Table 2.6

Nucleotide Excision Repair

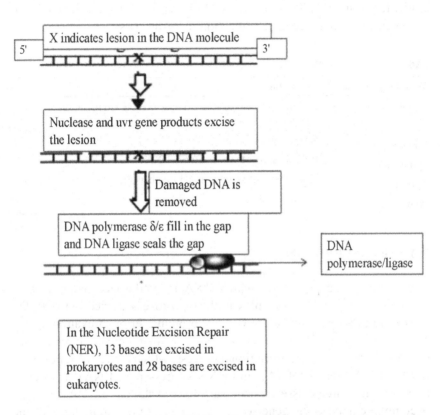

FIGURE 2.12   In DNA repair during DNA replication, there are chances of incorporation of wrong nucleotide by the polymerase, to correct, another property, that is, exonuclease activity of DNA polymerase I come into play which excises the wrong nucleotide and fills the gap thus produced by other nucleotides also known as mismatch repair.

All polymerase except beta has 3′→5′ exonuclease activity. In contrast to bacterial replication which starts at one point, there are many sites at which replication initiates which are known as replicons. This is because eukaryotic DNA is much larger than the prokaryotic and the entire length has to be duplicated within stipulated time. Another level of regulation of replication occurs in eukaryote where the time at which a region of chromosome undergoes replication in a particular cell type is programmed and predicted. Study of replication in eukaryotes was started in yeast cell which contains autonomous replicating sequences (ARS) which have the ability to promote replication of the DNA in which they are contained, and have helped us to understand replication in eukaryotes.

TABLE 2.6   Various DNA Polymerases of Eukaryotes.

DNA polymerase	Function
α(alpha)	Tightly associated with primase involved in initiation of primer synthesis
β(beta)	DNA repair
γ(gamma)	Replication of mitochondrial DNA
δ(delta)	Assembly of lagging strand fragments
ε(epsilon)	Undefined

2.4   TRANSCRIPTION

Transcription is a process in which RNA is synthesized using a DNA template. This is a very complex multistep process carried out with the aid of multiple enzymes and accessory proteins. Prior to examining the process of transcription it is necessary to know that there are three classes of RNA synthesize by transcriptional process, which are rRNA (ribosomal RNA), tRNA (transfer RNA), and mRNA (messenger RNA). Transcription of all the three classes is necessary for translation. mRNA molecule is complimentary to the gene sequence of one of the double helix of the two strands. tRNA has the anticodon against each triplet codon of mRNA. Thus, the function of tRNA is to bring the appropriate amino acid to the site of protein synthesis that is the ribosomes. The intermediate role of RNA between DNA and protein was elucidated by observing the following events:

1.   DNA remains within the nucleus whereas protein synthesis is a cytosolic event associated with ribosomes present in the cytoplasm.

There is no evidence to show direct participation of DNA in protein synthesis.

2. Synthesis of RNA takes place in nucleus of eukaryotes, which is also the site for DNA and has similar chemical nature as DNA.

3. RNA after synthesis in nucleus migrates to the cytoplasm where protein synthesis occurs.

4. It has been observed that amount of protein in a cell is usually proportional to the amount of RNA.

Thus, it is clear from the above mentioned points that genetic information stored in DNA is decoded in the form of protein with RNA as intermediate. The concept of a mRNA made using a DNA template to direct protein synthesis was proposed by Francois Jacob and Jacques Monod in 1961when they were working on model for gene regulation in bacteria. This has firmly been establishing now that RNA indeed acts as an intermediate between DNA and protein.

2.4.1   RNA SYNTHESIS BY RNA POLYMERASE

RNA polymerase was isolated by Samuel Weiss in 1959 from rat liver. RNA polymerase has the same substrate requirement as that of DNA polymerase with the exception of substrate nucleotide containing ribose sugar and not deoxyribose sugar. Another marked difference is the nonrequirement of primer for initiating the process.

$$n(NTP) \xrightarrow{\text{DNA enzyme}} (NMP)n + n(PPi)$$

This equation clearly demonstrates that nucleoside triphosphates (NTPs) are substrate for the enzyme to catalyze the polymerization of nucleoside monophosphates (NMPs) into a polynucleotide chain (NMP)n. nucleotides are linked together by 5'–3' phosphodiester bond. The whole reaction is summarized below.

$$(NMP)_n + NTP \xrightarrow{\text{DNA enzyme}} (NMP)_{n+1} + PPi$$

2.4.2 TRANSCRIPTION: PROKARYOTE

RNA polymerase of *E. coli* has been extensively studied and characterized (Table 2.7). It was shown to have many subunits designated as α, β, β', and σ. The active form or the holoenzyme contains subunits α_2, β, β', and σ having molecular weight of almost 500000 Da.

TABLE 2.7 Subunit Structure of RNA Polymerase of *E. coli.*

Subunit	Molecular weight	Number per polymerase	
Alpha	41,000	2	Core enzyme
Beta	155,000	1	
Beta Prime	165,000	1	Holoenzyme
Sigma	86,000	1	

In contrast to the eukaryotic cell, a prokaryotic cell contains single type of RNA polymerase composed of five subunits which are tightly associated to form core enzyme and a σ (sigma) factor which gives specificity to the enzyme (Fig. 2.13). In the absence of σ factor RNA molecules formed *in vitro* are not the same as that produced *in vivo*. σ factor increases the affinity of enzyme to the promoter site of DNA. The promoter in bacterial cell is located in upstream region of DNA strand preceding the site of initiation of transcription. There are two upstream stretches present, one at -35 position and other at -10 from the initiation site. The -10 region is known as *Pribnow box* after its discoverer (Fig. 2.14).

Bacterial cells have a variety of σ factor which recognizes different variations of -35 sequence which facilitates the selection of a particular battery of genes to be transcribed under certain condition, for example, when a bacterial cell is subjected to stress in the form of temperature. Under such stress condition new σ factor is synthesized which has the capability to recognize a different promoter sequence leading to transcription of heat shock genes. The second consensus sequence at -10 position or the Pribnow box occurs as a variation of sequence TATAAT (refer to Fig. 2.14). Mutation in -35 and -10 region affects transcription rate, thus the promoter region also controls the rate of transcription besides being the recognition sequences. As it is clear that transcription is initiated at specific points in chromosome, termination of this process also occurs at specific nucleotide sequence. A protein called the *rho* factor is sometimes required for termination of transcription process, but in most cases, the process is terminated and RNA chain is released without any additional factors.

Rho-dependent termination: such termination sites two stretches of G–C pairs arranged in inverted repeat manner which is followed by a string of adenines in the strand which has been transcribed. Due to this symmetry, the 3'end of the nascent RNA acquires a hairpin loop structure (Fig. 2.13) which stops the further activity of RNA polymerase. This mechanism also supports the importance of secondary structure of RNA. The adenine containing DNA region facilitates the release of RNA because base pairing between U in RNA and A in DNA strand are weak.

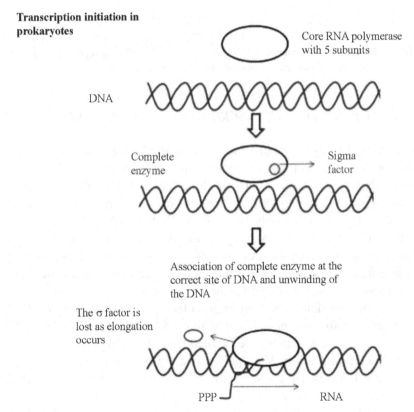

Transcription initiation in prokaryotes

Core RNA polymerase with 5 subunits

DNA

Complete enzyme

Sigma factor

Association of complete enzyme at the correct site of DNA and unwinding of the DNA

The σ factor is lost as elongation occurs

PPP

RNA

FIGURE 2.13   *E. coli* DNA dependent RNA polymerase involved in the transcription of mRNA and demonstration of rho-dependent termination of transcription.

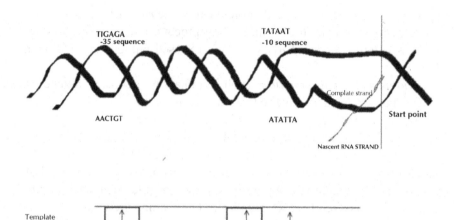

FIGURE 2.14   Promoter region in DNA of bacterium *E. coli.* The regulatory region is located at −35 and −10 base pairs from the transcription initiation site.

2.4.3   EUKARYOTIC TRANSCRIPTION AND RNA PROCESSING

Transcription is a process in which various RNAs are synthesized using DNA as a template. This process requires the action of specialized enzymes called RNA polymerases. There are three types of RNA that is mRNA, tRNA, and rRNA having specific function and location. Eukaryotic cells have three types of RNA polymerase for mRNA, rRNA, and tRNA.

Prokaryotic cell contains only single type of RNA polymerase as compared to a eukaryotic cell which has three types of polymerase starting from the simplest form that is the yeast cell. Eukaryotic RNA polymerases contain 8–14 polypeptides or subunits apart from various auxiliary proteins known as transcription factors. The three types of RNA are distinguishable on the basis of inhibition by α-amanitin (Table 2.8). α-amanitin is a toxic octapeptide isolated from *Amanita phalloides* which is a poisonous mushroom. RNA polymerase II is the most sensitive among the three polymerases to the inhibition by α-amanitin as compared to RNA polymerase I which is not sensitive and polymerase III which is sensitive to a lesser degree. Consumption of this poisonous mushroom leads to lack of production of mRNA necessary to direct protein synthesis. All the three

RNAs are derived from a longer primary transcript or pre RNA which contains nucleotides equivalent to the full length of DNA which has been transcribed. Pre RNA is processed into smaller functional RNA by undergoing several processing steps. Processing of RNA requires a number of small RNAs called small nuclear RNAs (sn RNAs) which are functional in nucleus (Table 2.9).

TABLE 2.8 RNA Polymerases to Synthesize RNAs in Eukaryotes.

RNA	RNA Polymerase	Inhibition by α-amanitin
rRNA (28S, 18S, and 5.8S)	I	No effect
mRNA	II	Shows maximum inhibition
tRNA and 5S rRNA	III	Inhibition observed to a lesser degree

There are different snRNAs (Table 2.9) involved in processing of RNAs. The three types of RNA have been described in subsequent part of this chapter.

TABLE 2.9 Small Nuclear RNAs.

RNA	Function Polymerase	RNA
Nucleoplasm	Pre mRNA splicing	II
U1	Pre mRNA splicing	II
U2	Pre mRNA splicing	II
U4	Pre mRNA splicing	II
U5	Pre mRNA splicing	II
U6	Pre mRNA splicing	III
U7	Histone pre mRNA splicing	II
U11	Unknown	II
U12	Unknown II	
7SK	Unknown	III
8-2	Pre tRNA processing	III
Nucleolus		
U3	Pre rRNA processing	II
U8	Unknown	II
U13	Unknown	II
2-Jul	Unknown	III

2.4.3.1   RIBOSOMAL RNAs

Ribosomes are site for protein synthesis. Each eukaryotic cell contains multiple ribosomes which contain several rRNA along with ribosomal proteins. Multiple repeats of DNA encoding rRNA is present in the genome which is called r-DNA. When the cell is in a nondividing state that is in interphase the cluster of r-DNA collects at a site called nucleolus which is a ribosome producing organelle. The nucleolus disappears during mitosis. Chromosomal region containing r-DNA is called nucleolar organizers. r-DNA is transcribed as a long transcript (pre rRNA) which is cleaved by number of enzymes to form a functionally active transcript. Another level of processing is the addition of methyl group to the transcript where methylated part is a part of final transcript and nonmethylated part is destroyed during processing. Processing of pre rRNA also involves association with ribonucleoprotein (RNP). Eukaryotes have four distinct type of ribosomal RNA, out of which three are in the large subunit and one is in small subunit. In case of higher eukaryotes such as humans, the large subunit is composed of 28S, 5.8S, and 5S molecule, whereas the small subunit contains 18S. 28S, 5.8S, and 18S are derived from a single transcript called the pre-RNA and 5S RNA is synthesized from separate precursor outside the nucleolus. Mature rRNA is formed after enzymatic cleavage of pre RNA and methylation. For processing of pre-RNA association with RNP particle occurs. The RNP particles get assembled in the cytoplasm then migrate to nucleus and lastly to nucleolus which also acts as a site for assembly of two ribosomal subunits. 5S RNA which is present as a part of large ribosomal subunit is transported separately to the nucleolus to join other components to form ribosomal subunit (Fig. 2.15a).

2.4.3.2   TRANSFER RNA

There are approximately 60 different types of tRNA in a eukaryotic cell. tRNA is transcribed by RNA polymerase III in a precursor form which is larger than the mature form. All mature tRNA have the triplet sequence CCA at their 3′ end (Fig. 2.15b) which is added after the tRNA has been processed. The function of the tRNAs is to bring the specific amino acid to the site of protein synthesis that is at the ribosomes with the aid of anticodon present on it. The anticodon facilitates in recognition of amino acids.

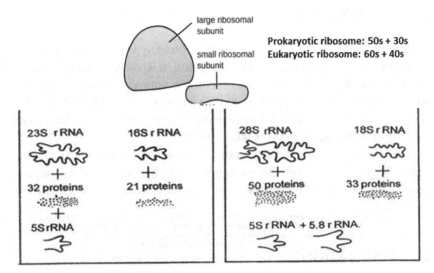

FIGURE 2.15a   Generalized structure of ribosomes in prokaryotic and eukaryotic cell.

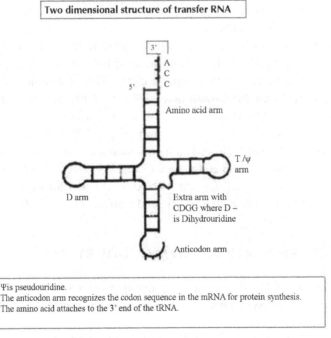

FIGURE 2.15b   Structure of tRNA is different from mRNA and rRNA with having secondary structure making cloverleaf structure.

2.4.3.3 MESSENGER RNA

Messenger (m) RNA is the template for protein synthesis. They are transcribed by RNA polymerase II. In eukaryotes, m RNA is posttranscriptionally modified by several processes including removal of exons, capping, and polyadenylation before going to the site of protein synthesis that is the ribosomes. mRNA is the link between the genetic information and the end product of this sequence that is the protein.

Therefore, it is clear that the process of transcription and translation are linked by mRNA. Both the above mentioned processes are highly orchestrated with the aid of multiple accessory proteins as well as rRNA, tRNA, and mRNA.

2.5 TRANSLATION

Products of transcription process that is mRNA, rRNA, and tRNA are involved in the process of translation which in the synthesis of protein. Translation is the process through which the information stored in genes is made use of for all cellular activities.

This process can be summarized in simple terms as the polymerization reaction occurring in the biological system where the amino acids are linked together to form a polypeptide chain. This is a multistep process (Fig. 2.16) in which the initiation is associated with the feasibility of the process. The initiating codon is methionine (Fig. 2.17). Effective termination of this process is also of utmost importance, this is achieved with the help of certain codons referred to as the stop codons which do not code for any amino acid. These codons are (Table 2.1) UAG, UAA, and UGA. These codons do not call for any tRNA. There is a lot of difference in translational process in a prokaryotic and a eukaryotic cell (Table 2.10).

2.6 GENE REGULATION AND REGULATORY RNA

A protein's cellular concentration is decided by a very fine balance of several processes, each having several potential points of regulation:

1) synthesis of the primary RNA transcript (transcription),
2) posttranscriptional modification of mRNA,

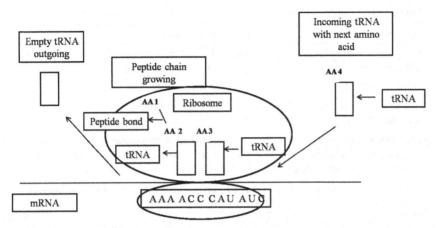

FIGURE 2.16 Pictorial representation of translation process which includes binding of mRNA to the small subunit of ribosomes thus initiating the process.

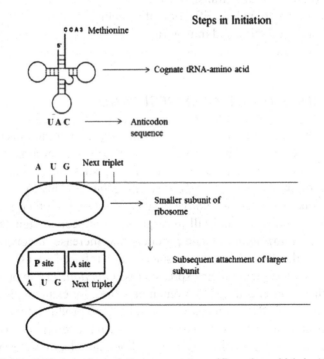

FIGURE 2.17 Initiation of translation occurs at specific codon which is AUG which codes for methionine. The amino acid methionine is brought to the site by the cognate tRNA at P site of ribosome. Subsequent amino acid is brought at A site and peptide bond is formed between the two amino acids.

TABLE 2.10   Features Differing in Prokaryote and Eukaryotic Translation.

Eukaryote	Prokaryote
Translation is a complex process occurring on larger ribosomes	Translation is not so complex and the size of ribosome smaller
mRNA long lived	mRNA sort lived
5′end of mRNA capped with 7-methyl-guanosine (7mG)	no capping observed
Most mRNA contain a short recognition sequence surrounding the initiating AUG codon, known as Kozac sequence	Shine-Dalgarno sequence present to facilitate binding of mRNA to small subunit of ribosome
Amino acid formylmethionine not required for initiation	Formylmethionine absolute requirement for initiation

3) messenger RNA degradation,
4) protein synthesis (translation),
5) posttranslational modification of proteins,
6) protein targeting and transport,
7) protein degradation.

2.6.1 PRINCIPLES OF GENE REGULATION

Genes for products that are required at all times, such as those for the enzymes of central metabolic pathways, are expressed at a more or less constant level in virtually every cell of a species or organism. Such genes are often referred to as *housekeeping genes.* Unvarying expression of a gene is called *constitutive gene expression.* For other gene products, cellular levels rise and fall in response to molecular signals; this is *regulated gene expression.* Gene products that increase in concentration under particular molecular circumstances are referred to as *inducible*; the process of increasing their expression is *induction.* The expression of many of the genes encoding DNA repair enzymes, for example, is induced by high levels of DNA damage. Conversely, gene products that decrease in concentration in response to a molecular signal are referred to as *repressible,* and the process is called *repression.* For example, in bacteria, ample supplies of tryptophan lead to repression of the genes for the enzymes that catalyze tryptophan biosynthesis. Transcription is mediated and regulated by protein–DNA interactions, especially those involving the protein components of RNA polymerase. We first consider how the activity of RNA polymerase is regulated, and proceed to a general description of the proteins participating in this process. We then examine the molecular basis for the recognition of specific DNA sequences by DNA-binding proteins.

At least three types of proteins regulate transcription initiation by RNA polymerase:

a) *Specificity factors* that alter the specificity of RNA polymerase for a given promoter or set of promoters.
b) *Repressors* involved in impeding the access of RNA polymerase to the promoter.
c) *Activators* which enhance the RNA polymerase–promoter interaction.

The σ subunit of the *E. coli* RNA polymerase holoenzyme is a specificity factor that mediates promoter recognition and binding. Most *E. coli* promoters are recognized by a single σ subunit (molecular weight 70,000) σ 70. Under some conditions, some of the σ70 subunits are replaced by another specificity factor. One case arises when the bacteria are subjected to heat stress, leading to the replacement of σ 70 by σ 32 (molecular weight 32,000). When bound to σ32, RNA polymerase is directed to a specialized set of promoters with a different consensus sequence. These promoters control the expression of a set of genes that encode the heat-shock response proteins. Thus, by making changes in the binding affinity of the polymerase that direct it to different promoters, a set of genes involved in related processes is coordinately regulated. In eukaryotic cells, some of the general transcription factors, in particular, the TATA-binding protein (TBP) may be considered specificity factors. Repressors bind to specific sites on the DNA. In prokaryotic cells, such binding sites, called *operators,* are generally near a promoter similar to RNA polymerase binding (Fig. 2.18). In prokaryotes, transcription rate is regulated by three mechanisms listed below:

a) by DNA binding protein,
b) by attenuation,
c) by exchange of sigma factor part of RNA polymerase.

For utilization of lactose which is a disaccharide, two major proteins are required galactose permease which mediates its entry into the cell and β-galactosidase which catalyes the hydrolytic cleavage of lactose into glucose and galactose. Expression of β-galactosidase increases multifold in the presence of lactose otherwise its level remains very low. Thus, the lactose or lac operon is inducible and is subject to catabolic repression. There the mechanism of regulation is under the control of DNA binding proteins, the lac repressor. Lac repressor is not a part of lac operon but only linked to it and its sole function is to mediate control by induction. The lac repressor forms a tetramer and binds with very high affinity and specificity to the operator site present between the promoter and structural genes and inhibits the transcription of structural genes in the absence of the substrate lactose. Lactose when present in the growth medium binds to the repressor and brings about the induction of lac operon.

In *E. coli*, the five contiguous *trp* genes encode enzymes that synthesize the amino acid tryptophan. These genes are expressed efficiently only when tryptophan is limiting (Fig. 2.18). The genes are controlled by a repressor, just as the *lac* genes are, although in this case, it is the *absence* of its ligand (tryptophan) that relieves repression. Even after RNA polymerase has initiated a *trp* mRNA molecule, however, it does not always complete the full transcript. As with riboswitches, the decision to make a complete transcript is controlled by attenuation; in this case, most transcripts are terminated prematurely, before they include even the first *trp* gene (*trpE*). But attenuation is overcome if tryptophan levels are low in the cell; when tryptophan is limiting, polymerase does not terminate and instead transcribes all of the *trp* genes. Whether or not attenuation occurs depends on the ability of RNAs to form alternative secondary structures, just as it did with the riboswitches. In this case, however, the choice between alternative structures formed by the leader RNA is not controlled by binding of ligand directly to that RNA; instead, the choice of alternatives relies on the coupling of transcription and translation in bacteria. The sequence of the 5'end of *trp* operon mRNA includes a 161-nucleotide leader sequence upstream of the first codon of *trpE* (Fig. 2.19). Near the end of this leader sequence, and before *trpE*, is a transcription terminator, composed of a characteristic hairpin loop in the RNA (made from sequences in regions 3 and 4 (Fig. 2.19), followed by eight uridine residues (Fig. 2.19). Transcription usually stops after this terminator, yielding leader RNA 139 nucleotides long. This is the RNA product seen in the presence of high levels of tryptophan. The *trp* operon is controlled by repression and attenuation, providing a two-stage response to progressively more stringent tryptophan starvation. But attenuation alone can provide robust regulation: other amino acid operons such as *leu* and *his* rely entirely on attenuation for their control. In the case of the leucine operon, its leader peptide has four adjacent leucine codons, and the histidine operon leader peptide has seven histidine codons in a row.

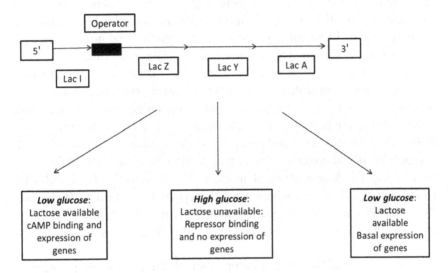

FIGURE 2.18 An example of inducible operon is Lac operon which is controlled with the help of a repressor protein.

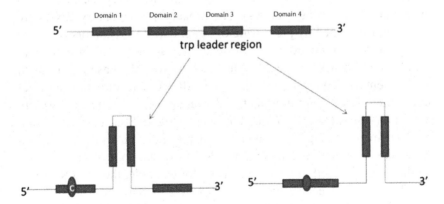

FIGURE 2.19 The *trp* operon.

2.6.2 REGULATORY RNA

There are many evidences which suggest that RNA functions not only act as a messenger between DNA and protein but also plays a very important role in the regulation of genome organization and gene expression. Genome organization is very complex and elaborate in higher organisms as compared to simple eukaryotes. For the formation and maintenance of such complex structure, RNA comes into play. This additional function of RNA has earned them the nomenclature of regulatory RNA. Regulatory RNA operates at multiple levels; that is at transcription level, translation level as well as it plays an important part in the epigenetic processes that control differentiation and development thus suggesting a central role of RNA in human evolution and ontogeny. RNA regulates with the aid of its various forms and methods, which may lead to silencing or total destruction of any RNA or protein of interest. The discovery of microRNAs (miRNAs) in eukaryotes which was first reported in the early 1990s, and then the discovery of the phenomenon known as RNA interference in the late 1990s has led to the emergence of this field known as regulatory RNA.

Apart from miRNA another group of small RNA molecules in bacteria that regulate translation and mRNA degradation have been discovered which have functional similarity to RNAs that regulate gene expression in eukaryotes—the small interfering (si) and miRNAs. The bacterial small RNAs commonly known as called *sRNAs* are much larger (80–110 nucleotides) than those regulatory RNAs present in eukaryotic cell (which range from 21 to 30 nucleotides). The prokaryotic regulatory RNAs are not generally formed by processing of larger dsRNA precursors (as those eukaryotic RNA regulators are); instead, they are encoded in their final form by small genes (Fig. 2.20). Most sRNAs work by base pairing with complementary sequences within target mRNAs and directing destruction of the mRNA, inhibiting its translation or even in some cases *stimulating* translation (as in case of riboswitches Fig. 2.24). Small RNAs are involved in regulating the replication of plasmids, as well as involved in regulation of gene expression. The RNAs which control transcription are the ones involved in gene regulation, for example, the 6S RNA of *E. coli*. The 6S RNA accumulates at high levels in stationary phase when the nutrient is depleted from the growth medium. This RNA binds to the σ^{70} subunit of RNA polymerase and downregulates transcription from many σ^{70} promoters. During stationary phase, an alternative σ factor, σS, is made. This σ competes with σ^{70} for the core region of polymerase and leads to expression of stress-related genes required for survival during the stationary phase. Thus, 6S RNA is able to control the expression of stress-related genes.

The sRNA of eukaryotes bind to its target mRNA and in most cases, this binding is aided by the bacterial protein Hfq which is a chaperone. Another well-studied sRNA from *E. coli* is the 81-nucleotide RybB RNA. This sRNA binds several target mRNAs and triggers their destruction by the nuclease RNase E (which acts on double strand of RNA). RybB mostly targets mRNA which encodes iron storage proteins. Free iron is required by the cell under certain circumstances, but high levels are toxic thus RybB is involved in the regulation of free iron level. This is done by controlling the levels of iron storage proteins that is their transcription. RybB is expressed from a promoter recognized by a special σ factor called σE (which is similar to σS, a stress response σ factor described earlier). Apart from the aforementioned function of RybB, expression of the gene encoding σE is itself regulated by RybB, and so this sRNA is also a part of an autonegative regulatory loop for σE.

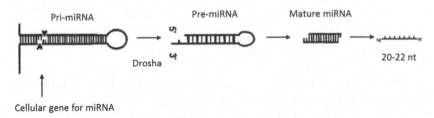

FIGURE 2.20 Synthesis of miRNA *in vivo* which clearly shows the synthesis of large precursor called the pri and pre miRNA.

Another mode of regulation is by riboswitches that are involved in the control of metabolic operons and attenuation in biosynthetic operons (Fig. 2.20). Riboswitches are the regulatory elements usually found within the 5-untranslated region of the genes which they control. The control can be exerted at both transcription and translation level. Each riboswitch is made up of two components: the aptamer and the expression platform (Fig. 2.24). The aptamer binds to the small-molecule ligand and then undergoes a conformational change, which ultimately causes a change in the secondary structure of the adjoining expression platform. These conformational changes alter expression of the associated gene by either terminating transcription or inhibiting the initiation of translation.

2.6.2.1 RNA INTERFERENCE, MICRO (mi) RNA AND siRNA

miRNAs are a class of small, endogenous RNAs which are of 21–25 nucleotides in length. They target specific mRNAs to be degraded or translationally repressed for regulation. Recent scientific advances have revealed the synthesis pathways and the regulatory mechanisms of miRNAs in animals and plants (Fig. 2.21).

Initial notions regarding miRNA and siRNA were:

1) miRNAs were viewed as endogenous and purposefully expressed products of an organism's own genome, whereas siRNAs were thought to be primarily exogenous in origin, derived directly from the virus, transposon, or transgene trigger.
2) miRNAs appeared to be processed from stem-loop precursors with incomplete double-stranded character, whereas siRNAs were found to be excised from long, fully complementary double-stranded RNAs (dsRNAs).

Despite these differences between miRNA and siRNA, similarities in their size and sequence-specific inhibitory functions boldly suggested the relatedness in their biogenesis and mechanism. The similarities as observed for both the classes of small RNAs were as follows:

1) They are similar in dependence upon the same two families of proteins: Dicer enzymes to excise them from their precursors.
2) There is a requirement of Ago proteins to support their silencing effector functions.

Thus, these three sets of macromolecules—Dicers, Agos, and 21–23 nucleotides long duplex-derived RNAs—became recognized as the signature components of RNA silencing.

The first small RNA, *lin-4*, was discovered in 1993 through a genetic screening in nematodes. Later in the same year, the regulation of *lin-14* by *lin-4* was discovered, which demonstrated the regulatory function of small RNAs[1] and [2]. The shorter *lin-4*RNA is now recognized as the origin of an abundant class of small regulatory RNAs, known as miRNAs. miRNAs are known to have functions in multiple processes which include immunity, hematopoietic differentiation, cell proliferation, cell death, etc., to name[3].

2.6.2.1.1 Mechanism of Silencing

miRNAs guide miRISC (RNA induced silencing complex) to specifically recognize messenger RNA (mRNA) and downregulate gene expression by one of the two posttranscriptional mechanisms:

1) translational repression and (2) mRNA cleavage (Figs. 2.21–2.23).

The regulatory mechanism is controlled and determined by the degree of miRNA–mRNA complementarity. The high degree of complementarity enables the Ago-catalyzed degradation of target mRNA sequences through the mRNA cleavage mechanism process. In contrast, a central mismatch omits degradation and facilitates the translational repression mechanism. A number of miRNA have been identified in plants and animals playing a very significant role (Table 2.11).

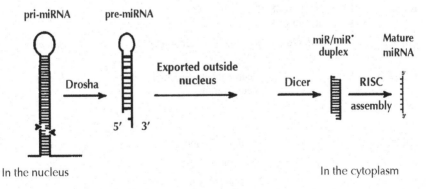

FIGURE 2.21 miRNA is transcribed in nucleus in the form of pri miRNA.

TABLE 2.11   miRNA Present in Animals and Their Function. [1, 27–51]

miRNAs	Target gene	Biological functions	Species
Bantam	HID	Cell death and proliferation	*D. melanogaster*
let-7	*lin-41*, HBL-1	Regulation of developmental timing	*C. elegans*
lin-4	*lin-14*, lin-28	Physiological condition and developmental timing	*C. elegans*
lsy-6	COG-1	Neuronal cell fate and developmental timing	*C. elegans*
miR-1	HAND 2	Cardiomyocyte differentiation and proliferation	*Mus musculus*
miR-7	Notch targets	Notch signaling	*D. melanogaster*
miR-14	Caspase?	Cell death and proliferation	*D. melanogaster*
miR-15a, miR-16-1	Bcl$_2$	Down-regulated in B cell chronic lymphocyte leukemia	
miR-16	Several	AU-rich element-mediated mRNA instability	*Homo sapiens*
miR-17-92	c-Myc, E2F1	Upregulated in B-cell lymphoma	*H. sapiens*
miR-32	Retrovirus PFV1	Antiviral defense	*H. sapiens*
miR-143	ERK5	Adipocyte differentiation	
miR-143, miR-145	Unknown	Downregulated in colonic adenocarcinoma	*H. sapiens*
miR-146	c-Myc, ROCK1	Development and function of immune system	*H. sapiens*
miR-155	PU-1, c-Maf	T-cell development and in innate immunity	Mouse
miR-181	Unknown	Regulation of hematopoietic cell fate	*M. musculus*
miR-196	HOXA7, HOXB8, HOXC8, HOXD8	Development?	*M. musculus*
miR-223	NFI-A, Mef2c	Regulation of granulocytic maturation	*H. sapiens*
miR-273	DIE-1	Neuronal cell fate and developmental timing	*C. elegans*
miR-372, miR-373	LATS2		
miR-375	Myotrophin	Insulin secretions	*M. musculus*
miR-430	?	Brain morphogenesis	*D. rerio*
SVmiRNAs	SV40 viral mRNAs	Susceptibility to cytotoxic T cells	

2.6.2.1.2 Translation Repression

The exact mechanism for the repression of target mRNA translation by miRISC is still unknown. Whether repression occurs at the translational initiation or posttranslational level still needs to be determined. However, the current model suggests that the eIF4F complex is involved in translational initiation. The subunits of the eIF4F complex include eIF4A, eIF4E, and eIF4G. The mRNA 5′ terminal cap is recognized by eIF4E and thus starts the initiation process. eIF3, another initiation factor, interacts with eIF4G and contributes to the 40S ribosomal subunit assembly at the 5′end of the mRNA to enable the preinitiation complex. The elongation process is initiated by joining of the 60S ribosomal subunit at the AUG codon of the mRNA and the 40S preinitiation complex. eIF4G and eIF3 also interact with the polyA-binding protein PABP1. The mRNA molecule becomes circular as a result of this process, and the translation efficiency is thereby improved. In some viral mRNAs, the translation initiation process is facilitated without any initiation factors through internal ribosome sites (IRES), which require only a subset of the initiation factors[4].

Whether a miRNA inhibits translation through inhibition of initiation or elongation is typically determined by two sets of criteria. For the first option, the density gradient centrifugation technique is used to determine whether mRNAs are present in the complex mRNA–protein (mRNP) system (initiation inhibition), or in the form of large polysomes (elongation inhibition). The second criterion is tested by determining whether inhibited mRNAs containing IRES sequences are resistant to repression.[4,5] In testing this, some studies reported data supporting repressed initiation,[5–8] whereas others provide evidence for inhibition of the post-initiation processes.[9–11] However, none of the above criteria alone is sufficient to explain repressed initiation or inhibition of post-initiation processes. The existing discrepancies show that repression may occur either at the initiation step or at a later stage in the translation process.

In 2006, Petersen et al. proposed a possible mechanism through which miRISC may exert its action by repressing the elongation process. An inhibited mRNA can be associated with polysomes, but when the initiation process is rapidly blocked with hippuristanol, the ribosomes quickly become detached in a miRNA-dependent manner. On the basis of these results, it was suggested that miRISC promotes early ribosome dissociation from mRNAs. Recently, three different models have been proposed to explain the mechanism by which miRISC represses the initiation mechanism (Fig. 2.22). First, miRISCs were shown to compete with eIF4E for binding to the mRNA 5′ cap structure, which results in the failure of the translation initiation process.[7,12] However, some studies contradict this model and suggest that either GW182 or a downstream factor could be the eIF4E competitor[13]. The second model suggests that miRISC prevents the mRNA from circularizing, resulting in translation inhibition[14–17]. The C–C chemokine receptor 4-negative on TATA (CCR4–NOT) complex is composed of multiple proteins, namely

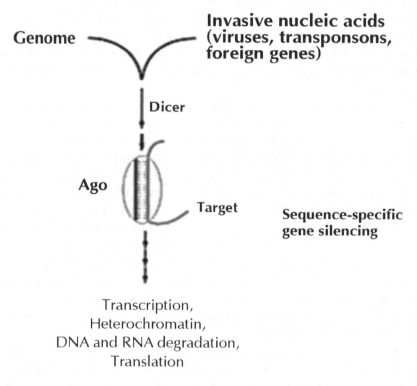

FIGURE 2.22 Core features of miRNA and siRNA silencing which can also act as a defense mechanism against invasive nucleic acid.

chemokine (C-C motif) receptor 4 (CCR4), chromatin assembly factor 1 subunit (CAF1), and NOT1–NOT5. These regulate gene expression and may be involved in miRISC translation inhibition[17–20]. The third model proposes that miRISC may inhibit the assembly of the 60S ribosomal subunit with the 40S preinitiation complex. In this process, the 40S ribosomes are attached to the targeted mRNA, but the 60S ribosomal subunit fails to join the 40S subunit, resulting in translation repression[21,22] (Fig. 2.23).

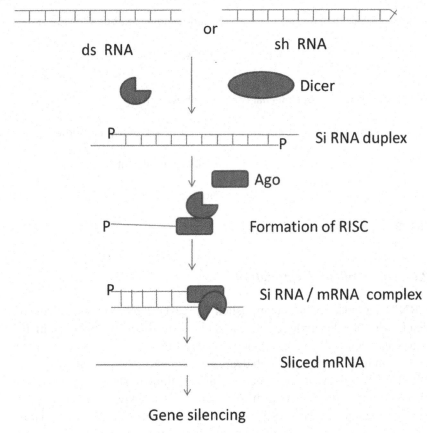

FIGURE 2.23   Mechanism of siRNA silencing.

Another possible mechanism of miRNA mediated translational repression is that miRNA/RISC may mediate translation repression through accumulation of target mRNAs in processing bodies (P-bodies)[23]. P-bodies lack any translation machinery, and thus, it is suggested that P-bodies containing mRNAs are not involved in the translation process[23]. The accumulation of mRNA in a miRNAs-dependent manner suggests that miRNAs are increasing the ribosome-free mRNA and cause translation repression.

A. Non-nucleolytic repression mechanism (thiM and btu B riboswiches)

Ligand + Ribiswitch (on state) $\rightleftarrows$ Ligand binding to riboswitch (off state)
$\downarrow$
No translation

B. Nucleolytic repression mechanism (lys C riboswitch)

Ligand + Riboswitch (on state) $\longrightarrow$ Binding action of degradasome
$\downarrow$
Riboswitch (off state)
$\downarrow$
No translation

FIGURE 2.24   Mechanism of action of riboswitches.

2.6.2.1.3   mRNA Degradation

mRNA degradation is accomplished by several mechanisms; they are deadenylation, decapping, and exonucleolytic digestion apart from the Ago-catalyzed mRNA degradation process. Previously, it has been shown that when miRNAs have a high degree of sequence complementarity, then target mRNA degradation processes are facilitated through Ago protein slicer activity. The fact that mRNAs are reduced with an abundance of miRNAs suggests that miRNAs are responsible for mRNA degradation processes.[14,15,24,25] mRNA degradation by miRNA requires Ago, GW182, and the cellular decapping and deadenylation machinery[13]. The exact process of target selection has yet to be determined. However, it has been shown that the number, type, and position of mismatches in the miRNA/

mRNA duplex play a critical role in the selection of the degradation or translational repression mechanisms.[26]

2.7   CLUSTERED REGULARLY INTERSPERSED SHORT PALINDROMIC REPEATS

Another aspect of innate immunity which deals with viral DNA or RNA in a prokaryotic cell has been worked out recently. Many bacterial and most archaeal genomes possess loci comprised regularly spaced repeats interspersed by other DNA sequences derived from viruses. These loci, now termed clustered regularly interspersed short palindromic repeats (CRISPRs) act as an innate immune system by incorporating fragments of viral DNA between the repeats, which are then transcribed and processed to produce small guide RNAs (linked to their effector complexes via the repeat sequence) that target and destroy viral DNA or RNA) (Fig. 2.25). This system has recently been adapted for RNA programmable sequence-specific genome manipulation in eukaryotes, including mammals, with extraordinary versatility including targeted gene excision and fusion, and modified CRISPRs capable of recruiting silencing and activating proteins to target loci. Moreover, the biological arms race continues, with bacterio-phages encoding their own CRISPR system to evade host innate immu-nity (for details, see Chapter 3 and 5). The CRISPR RNA associates with and guides bacterial molecular machinery to a matching target sequence in the invading virus. The molecular machinery cuts up and destroys the invading viral genome (Fig. 2.26).

CAS-CRISPR associated genes

The CRISPR immune system works to protect bacteria from repeated viral attack via three basic steps.

1. Adaptation—DNA from an invading virus is processed into short segments that are inserted into the CRISPR sequence as new spacers.
2. Production of CRISPR RNA—CRISPR repeats and spacers in the bacterial DNA undergo transcription, the process of copying DNA into RNA (ribonucleic acid). Unlike the double-chain helix structure of DNA, the resulting RNA is a single-chain molecule. This RNA chain is cut into short pieces called CRISPR RNAs.

3. Targeting—CRISPR RNAs guide bacterial molecular machinery to destroy the viral material. Because CRISPR RNA sequences are copied from the viral DNA sequences acquired during adaptation, they are exact matches to the viral genome and thus serve as excellent guides.

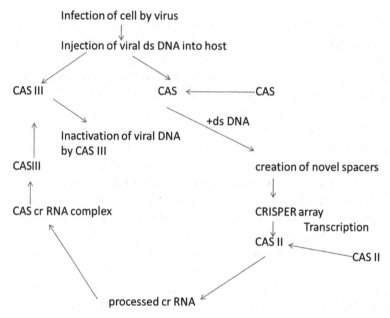

FIGURE 2.25   The steps of CRISPR-mediated immunity.

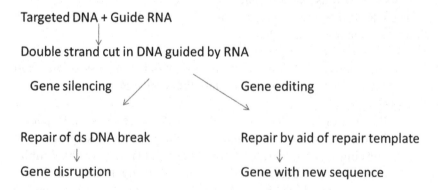

FIGURE 2.26   Gene silencing and editing with CRISPR.

2.8 MUTATION

A gene is a DNA segment which governs the process of transcription and translation, ultimately leading to its expression. Genes may occur individually on the chromosomes which undergo transcription and translation into a single type of protein. But at times, a cluster of genes encoding related function are transcribed from a single promoter forming a multigenic mRNA molecule, which is translated to form proteins whose number is equal to that of the genes in the cluster.

Events occurring during normal chromosomal replication or exposure to certain chemical and physical agent can lead to changes in the sequence of bases in genome, which can be termed as mutation. The agents responsible for bringing this change are called mutagens. There are various types of mutation events encountered by genome which may or may not prove lethal for the cell.

Mutation may be classified in various ways. First and foremost is spontaneous and induced mutation, where spontaneous mutation is a natural phenomenon with no specific mutagens or genes are associated with them. Their occurrence has greatly been observed during replication process. Induced mutation occurs due to the influence of various external factors such as radiations and chemicals. Induced mutation by X-ray was first demonstrated by Hermann J. Muller in *Drosophila*.

The second classification is based on the cells in which mutation is occurring, that is, somatic and gametic mutation. When mutation occurs in genome of somatic cells, it is not passed on to future generation and further if this event has occurred in a recessive allele it does not show any effect in the organism. But this has a marked effect if mutation has occurred in dominant allele or if it has occurred early in development when differentiation is still undergoing in the tissue and organ. Gametic mutation occurs in the gametes and this is carried further to the future generation. Such mutation if dominant shows the phenotypic effect in the first generation. Occurrence of autosomal recessive mutation in gametic cells may not be noticed for many generations and their phenotypic expression would occur by a chance mating which brings two copies in a homozygous condition.

Next classification is based on the stretch of genome being affected.

A) *Microleision*: mutation involving a change in only a single base pair of the cell's DNA.

(a) Frame-shift mutation; loss or gain of a base pair leading to change in reading frame of all codons of the gene or operon which may be at a distal end from the point of mutation. This type of mutation can be of +1 or −1 type due to addition or deletion of single base, causing the change in all the amino acids being incorporated in the growing peptide chain.

(b) *Base-pair substitution*; microlesion involving the change of one base pair (Adenine–Thymine, A–T) to another (Guanine–Cytosine, G–C). When a purine base is changed to another purine base or a pyrimidine to another pyrimidine base, it is called transition. Similarly, when a purine base is changed to a pyrimidine base or vice-versa it is called transversion.

1. Neutral—base pair substitution resulting in change of a codon to a redundant one encoding the same amino acid.

2. Missense—base pair substitution resulting in change of codon to a different amino acid. For example, sickle cell anemia is caused by a substitution in the beta-hemoglobin gene, which alters a single amino acid in the protein produced.

3. Nonsense—base pair substitution resulting in codon changed to become one of the stop codons resulting in premature termination of translation.

B) *Macroleision:* mutation involving a number of base pairs or even a number of genes which results in the generation of one or more than one junctions between DNA segments not present in the unmutated DNA. Such mutations have variable results as listed below:

1) Deletion: loss of gene function due to removal of a gene.

2) Duplication: results in generation of second copy of a gene.

3) Inversion: this type of mutation may affect several genes, due to inversion of DNA segment within a gene.

4) Insertion: when an inverted segment is inserted within a gene hindering its function.

5) Translocation: type of mutation in which a DNA segment is replicated and translocated to other sites in the genome, causing mutation at the point where translocated segment is found.

Besides these classifications various other types of mutation are also there, they are *nutritional or biochemical variations*. Inability to synthesize an amino acid or a vitamin by bacteria or fungus (auxotroph) is known as nutritional mutation, which has been put to great use for various studies. The experiment of Beadle and Tatum in which wild-type *Neurospora* was grown in a minimal medium and the spores were exposed to UV irradiation. The irradiated spores were cultured in supplemented medium which underwent meiosis and mitosis. The resulting spores were grown in minimal as well as supplemented media and cell samples were found to grow on minimal media supplemented with vitamin but not amino acids (auxotrophs) (Fig. 2.27).

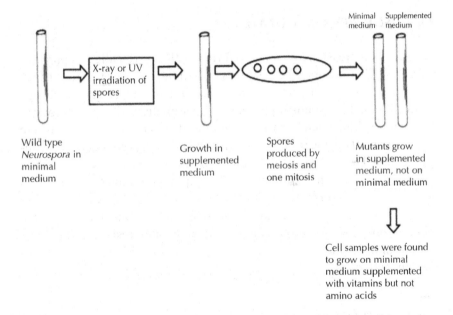

FIGURE 2.27 Experiment of Beadle and Tatum in which wild-type *Neurospora* was grown in a minimal medium and the spores were exposed to UV irradiation. The irradiated spores were cultured in supplemented medium which underwent meiosis and mitosis. The resulting spores were grown in minimal as well as supplemented media and cell samples were found to grow on minimal media supplemented with vitamin but not amino acids (auxotrophs).

Sickle cell anemia, Huntington disease in human can be classified as biochemical mutation which can also be lethal. Such mutation can be lethal. Another category is of *behavior mutation* seen in the changes of mating behavior of a fruit fly. Regulatory mutations are referred to those occurring in a regulatory gene which controls the transcription of another gene which can lead to activation or inactivation of the gene in question, as seen in many types of cancer where mutation in certain regulatory gene results in loss of cell cycle control. Lastly, there is *conditional mutation*, example of which is temperature sensitive mutation in various organisms. In such organisms, the mutant gene product functions properly at a permissive temperature but becomes functionless at nonpermissive or restrictive temperature.

2.8.1 SUPPRESSOR MUTATION

Mutation suppression by mutant form of tRNA is called suppressor mutation. This occurs when there is mutation in gene encoding tRNA which are called suppressors or suppressor mutation resulting in change in translational mechanism producing functionally active gene product. This can be illustrated with the example of mutation in *E. coli* where UAG is generated due to A/T to T/A transversion thus changing the wild-type codon UUG (codes for leucine) to amber nonsense codon UAG, which signals chain termination. The suppressor mutation here is a G/C to C/G transversion in gene encoding one of tyrosine tRNAs. This mutation changes the anticodon from AUG to AUC thus facilitating the recognition of amber codon and inserts a tyrosine residue at this site. Suppression is said to occur if the tyrosine containing residue is functionally active.

2.8.2 MUTAGENS

Various chemical or physical agents which cause or enhance the rate of mutation are known as mutagens (Table 2.12).

TABLE 2.12 Various Agents which Cause Mutation.

Name	Structure	Action
(1) Base analogue, e.g., 2-aminopurine	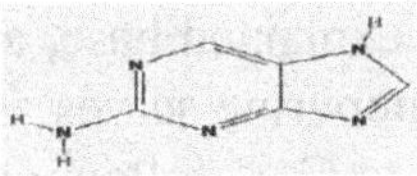	Incorporates into DNA
(2) Intercalating agent, e.g., ICR-191	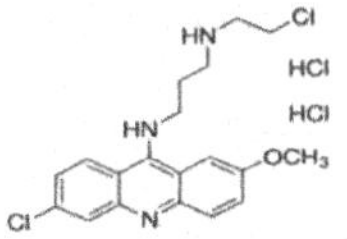	Causes frameshift mutation
(3) Alkylating agent, e.g., nitrosoguanidine	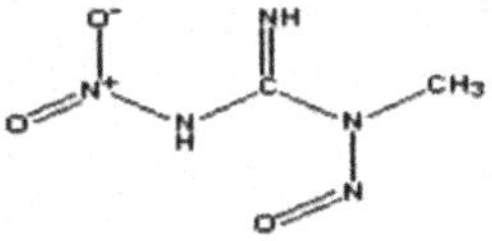	Alkylates purines and cause transition, transversion, and −1 frameshift mutation
(4) Other DNA modifiers, e.g., hydroxylamine	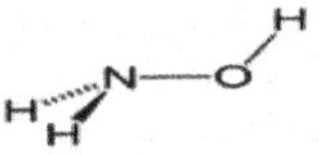	Hydroxylates 6 amino groups of cytosine resulting in G–C to A–T transitions

2.8.2.1 BASE ANALOGS

Molecules which act as mutagen by substituting for purines or pyrimidines during nucleic acid biosynthesis wherein normal condition adenine pairs with thymine and guanine pairs with cytosine (Fig. 2.28). For example, 5-bromouracil acts as a base analog for thymine.

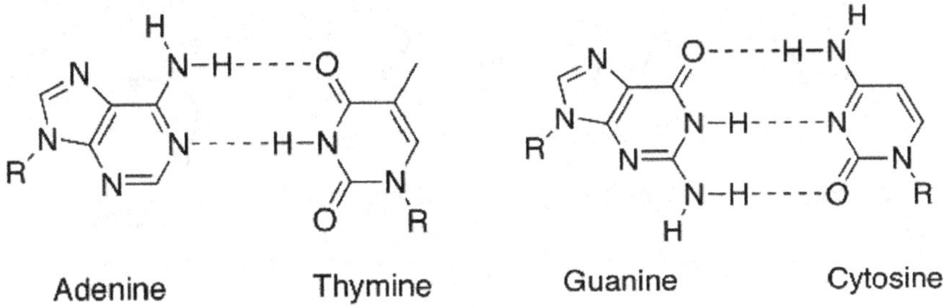

FIGURE 2.28 Standard base pairing by formation of hydrogen bond between adenine (amino) thymine (keto) and guanine (keto) cytosine (amino).

If 5-bromouracil is incorporated into DNA instead of thymine and a tautomeric shift to the enol form occurs, 5-bromouracil base pairs with guanine. Thus, after a round of replication cycle, an A=T is converted to G≡C. Apart from the above explained base, analog other analogs are also there, for example, 2-aminopurine serves as an analog for adenine causing transition from A=T is to G≡C following replication (Figs. 2.29 and 2.30).

Biochemical basis of mutating agents (Induced)

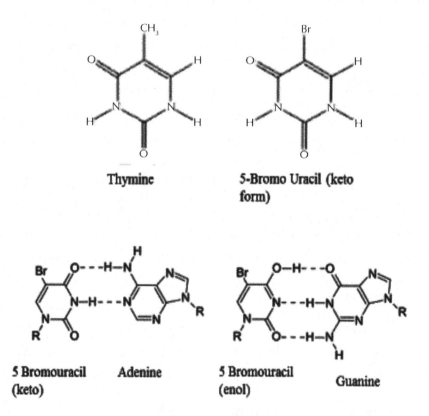

FIGURE 2.29   5-bromouracil which is a base analog of uracil forms hydrogen bond with adenine when present in keto form and with guanine when present in enol form. Incorporation of base analogs in DNA results in mutation.

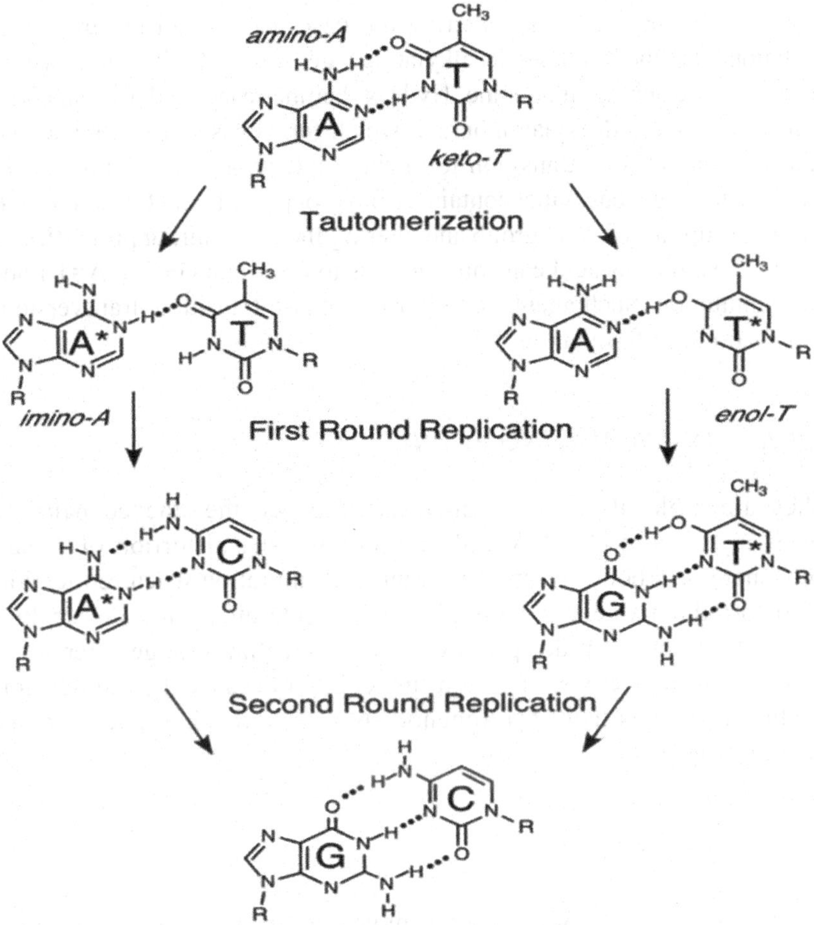

FIGURE 2.30 Conversion of AT base pair to GC base pair by action of base analog.

2.8.2.2 ALKYLATING AGENTS

These chemicals donate an alkyl group such as a methyl ($-CH_3$) or an ethyl group (CH_3-CH_2) to amino or keto groups in nucleotides. Example of alkylating agent used as a mutagen is ethyl methanesulfonate (EMS). EMS alkylates the keto group in the number 6 position of guanine and number 4 position of thymine, thus altering the base pairing causing transition mutation. Another mutagenic alkylating agent is 6-ethylguanine which acts as a base analog of adenine. Hydroxylamine reacts with cytosine and converts it to 6-hydroxylaminouracil which forms base pair with adenine causing $G \equiv C$ to $A=T$ transition following replication cycle. Nitrous acid reacts with all the bases that contain amino group (A, T, and C), converting amino group to hydroxyl group and altering the base pairing properties of them. Thus, nitrous acid can cause $A=T$ is to $G \equiv C$ and $G \equiv C$ to $A=T$ transition mutation. Such agents are known to cause transition, transversion, and -1 frameshift mutation.

2.8.2.3 INTERCALATING AGENTS

They are molecules which can insert between the stacked pairs of bases in the core of DNA molecule causing the distortion of sugar-phosphate backbone resulting in frameshift mutation when replication of distorted helix occurs. Examples of intercalating agents are acridine dyes, ethidium bromide, proflavine, etc. Acridine orange intercalate between purines or pyrimidines of intact DNA because their dimension is almost same as that of nitrogenous base pair causing deletions and insertions.

2.8.2.4 RADIATION

Many types of radiation also act as mutagen, for example, X-ray cause breaking of phosphodiester bonds, chromosomal break leading to formation of microlesions (Fig. 2.31). X-rays are short wavelength cosmic rays which are able to penetrate the tissue, causing ionization of the molecules.

Upon X-ray, penetration electrons are ejected from the atoms of the molecules transforming them into free radicals and reactive ions which can start a range of chemical reactions affecting the genetic material causing point mutation. Ultraviolet (UV) rays cause dimerization of thymine. Dimerization triggers a repair mechanism which cleaves the dimer thus exposing a short single-stranded region of DNA on the opposite strand (Figs. 2.31 and 2.32). Such single-stranded regions in the opposite strands of DNA are also formed if replication fork passes a dimer which has not been excised. These single-stranded regions induce the DNA repair mechanism called switch over something (SOS) system which is a fast acting but error-prone system filling the gaps in opposite strand thus introducing mutation following UV irradiation which may be lethal for the organism. If mutation inactivates the UV repair mechanism in an individual, they are always at a risk to develop epidermal malignancy on exposure to UV component of sunlight. The thymine dimer is cleaved by enzyme photolyase (Fig. 2.32) which is a light-dependent enzyme.

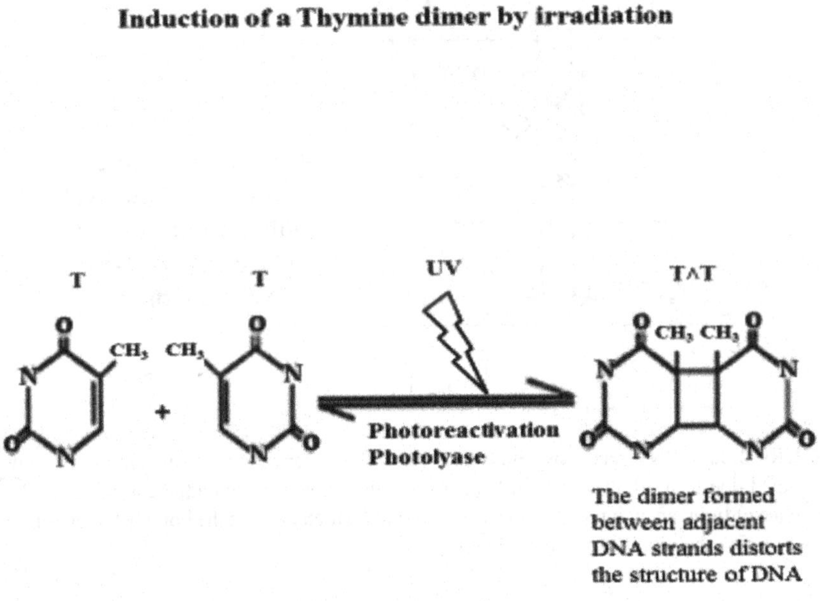

Induction of a Thymine dimer by irradiation

FIGURE 2.31 Thymine dimer formation takes place when a cell is irradiated with UV light. The formation of dimer results in the formation of gaps in the DNA which has to be repaired to avoid mutation.

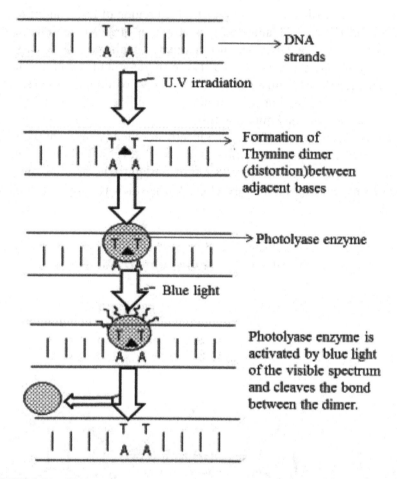

FIGURE 2.32   DNA repair by photoreactivation/light-dependent repair. Thymine dimer is formed due to irradiation, which leads to the distortion between adjacent bases. In the presence of light, an enzyme photolyase is activated which cleaves the bond between dimer thus bringing the repair of thymine dimer.

2.8.3 THE EFFECT OF MUTATION

DNA is present in all nucleated cells, thus providing a lot of sites for mutation to occur. Mutagenic effects which cannot be passed to offspring are called somatic mutation. Similarly, germ line mutations are the once occurring in reproductive cells, for example, egg cell and sperm cell and can be passed on to the offsprings. Thus, mutations which are carried forward to the next generation play an important role in evolutionary processes.

There are various outcome of germ line mutation as mentioned:

1) Mutation only occurs at genetic level and no phenotypic change is observed. This can be explained by the fact that the site at which mutation has occurred might be in a stretch of DNA with apparently no function, or the mutation has occurred in a protein coding region which is not affecting the amino acid sequence.
2) Mutation results in a small change in the phenotype.
3) Mutation results in a drastic change in phenotype, for example, development of resistance in insects against DDT.
4) Lethal mutations result in death of organism.

Certain DNA regions control other genes which determine when and where other genes are turned "on." If mutation occurs in such region it can greatly affect the genome with an impact multiplied to many folds. Many organisms have powerful control genes that determine how the body is laid out. For example, Hox genes are found in many animals (including flies and humans) and designate where the head goes and which regions of the body grow appendages. Such control genes help direct the building of body "units" such as segments, limbs, and eyes. So evolving a major change in basic body layout may not be so unlikely; it may simply require a change in a *Hox* gene and the favor of natural selection as seen in case of *Drosophila* where investigators have been able to sprout leg from head region.

2.8.3.1   SICKLE CELL ANEMIA

Sickle cell anemia is a genetic disease with severe symptoms which includes pain and anemia. The disease is caused by a mutation in the gene that helps makes hemoglobin. Hemoglobin is a protein that carries oxygen in red blood cells. This disease manifests in homozygous condition, that is, two copies of gene is required. Heterozygous individuals can act as carrier.

The mutations that cause sickle cell anemia have been extensively studied and demonstrate how the effects of mutations can be traced from the DNA level up to the level of the whole organism. The mutation has its effect on the DNA as well as protein/cellular level (Fig. 2.34). When red blood cells carrying mutant hemoglobin are deprived of oxygen, they become "sickle-shaped" instead of the usual round shape affecting and interrupting blood flow. Thus, this shape is unfavorable at high altitude where it can lead to symptom such as pain and fatigue in carriers of sickle cell. Contrary to this, sickle cell is favored in regions affected by malaria, because the causative organism does not survive in sickle-shaped RBCs.

Even if present in heterozygous condition certain effects is visible along with the carrier status.

Mutation in DNA sequence leads to substitution of T with A resulting in change of amino acid which is the change occurring at protein level. This lead to change in the structure of the protein leading to clumping of RBCs of affected organism.

2.8.3.2 MUTATIONS ARE RANDOM

Mutations can be beneficial, neutral, or harmful to the organism, but mutations do not "try" to supply what the organism "needs." Factors in the environment may influence the rate of mutation but are not generally thought to influence the direction of mutation. For example, exposure to harmful chemicals may increase the mutation rate, but will not cause more mutations that make the organism resistant to those chemicals. In this respect, mutations are random—whether a particular mutation happens or not is unrelated to how useful that mutation would be.

For example, in the United States where people have access to shampoos with chemicals that kill lice, we have a lot of lice that are resistant to those chemicals. Researchers have performed many experiments in this area. Though results can be interpreted in several ways, none unambiguously support directed mutation. Nevertheless, scientists are still doing research that provides evidence relevant to this issue.

In addition, experiments have made it clear that many mutations are in fact random, and did not occur because the organism was placed in a situation where the mutation would be useful. For example, if bacteria are exposed to an antibiotic, an increased prevalence of antibiotic-resistant cells would be observed. Esther and Joshua Lederberg determined that many of these mutations for antibiotic resistance existed in the population even before the population was exposed to the antibiotic and that exposure to the antibiotic did not cause those new resistant mutants to appear.

2.8.3.3   THE LEDERBERG EXPERIMENT

In 1952, Esther and Joshua Lederberg performed an experiment that helped to show that many mutations are random, not directed. In this experiment, they capitalized on the ease with which bacteria can be grown and maintained. Bacteria grow in isolated colonies on plates. These colonies can be reproduced from an original plate to new plates by "stamping" the original plate with a cloth and then stamping empty plates with the same cloth. Bacteria from each colony are picked up on the cloth and then deposited on the new plates by the cloth (Fig. 2.33).

1) Bacteria are spread out on a plate, called the "original plate."

(2) They are allowed to grow into several different colonies.

(3) This layout of colonies is stamped from the original plate onto a new plate that contains the antibiotic penicillin.

(4) Colonies X and Y on the stamped plate survive. They must carry a mutation for penicillin resistance.

(5) The question arose, "did the colonies on the new plate evolve antibiotic resistance because they were exposed to penicillin?" The answer is no. When the original plate is washed with penicillin, the same colonies (those in position X and Y) live—even though these colonies on the original plate have never encountered penicillin before.

FIGURE 2.33   Experiment to prove that preexposure to antibiotic is not required for the development of antibiotic resistance in bacteria.

Esther and Joshua hypothesized that antibiotic-resistant strains of bacteria surviving an application of antibiotics had the resistance before their exposure to the antibiotics, not as a result of the exposure. Their experimental setup is summarized below:

Normal red blood cells

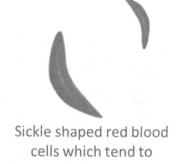

Sickle shaped red blood cells which tend to clump

FIGURE 2.34   Normal hemoglobin (left) and hemoglobin in sickle red blood cells (right) look different.

Therefore, the penicillin-resistant bacteria were there in the population before they encountered penicillin. They did not evolve resistance in response to exposure to the antibiotic (Fig. 2.33).

Plant–bacterial interactions in the rhizosphere are the determinants of plant health and soil fertility. Free-living soil bacteria beneficial to plant growth usually referred to as plant growth promoting rhizobacteria (PGPR), are capable of promoting plant growth by colonizing the plant root. PGPR are also termed plant health promoting rhizobacteria (PHPR) or nodule promoting rhizobacteria (NPR). These are associated with the rhizosphere, which is an important soil ecological environment for plant–microbe interactions. Symbiotic nitrogen-fixing bacteria include the cyanobacteria of the genera *Rhizobium, Bradyrhizobium, Azorhizobium, Allorhizobium, Sinorhizobium,* and *Mesorhizobium.* Free-living nitrogen-fixing bacteria or associative nitrogen fixers, for example, bacteria belonging to the species *Azospirillum, Enterobacter, Klebsiella,* and *Pseudomonas,* have been shown to attach to the root and efficiently colonize root surfaces. PGPR have the potential to contribute to sustainable plant growth promotion. Generally, PGPR function in three different ways: synthesizing particular compounds for the plants, facilitating the uptake of certain nutrients from the soil, and lessening or preventing the plants from diseases. Plant growth promotion and development can be facilitated both directly and indirectly. Indirect plant growth promotion includes the prevention of the deleterious effects of phytopathogenic organisms. This can be achieved by the production of siderophores, that is, small metal-binding molecules. Biological control of soil-borne plant pathogens and the synthesis of antibiotics have also been reported in several bacterial species. Another mechanism by which PGPR can inhibit phytopathogens is the production of hydrogen cyanide (HCN) and/or fungal cell wall degrading enzymes, for example, chitinase, and ß-1,3-glucanase. Direct plant growth promotion includes symbiotic and non-symbiotic PGPR which function through production of plant hormones such as auxins, cytokinins, gibberellins, ethylene, and abscisic acid. Production of indole-3-ethanol or indole-3-acetic acid (IAA), the compounds belonging to auxins, have been reported for several bacterial genera. Some PGPR function as a sink for 1-aminocyclopropane-1-carboxylate (ACC), the immediate precursor of ethylene in higher plants, by hydrolyzing it into α-ketobutyrate and ammonia, and in this way, promote root growth by lowering indigenous ethylene levels in the micro-rhizo environment. PGPR also help in solubilization of mineral phosphates and other nutrients, enhance resistance to stress, stabilize soil aggregates, and improve soil structure and organic matter content. PGPR retain more soil organic N, and other nutrients in the plant–soil system, thus reducing the need for fertilizer N and P and enhancing release of the nutrients.

KEYWORDS

- **replication**
- **transcription**
- **translation**
- **Okazaki fragments**
- **housekeeping genes**
- **mutation**
- **genetic code**
- **codon**

REFERENCES

1. Lee, R. C.; Feinbaum, R. L.; Ambros, V. The *C. elegans* Heterochronic Gene Lin-4encodes Small RNAs with Antisense Complementarity to Lin-14. *Cell* **1993**, *75*, 843–854.
2. Wightman, B.; Ha, I.; Ruvkun, G. Posttranscriptional Regulation of the Heterochronic Gene Lin-14 by Lin-4 Mediates Temporal Pattern Formation in *C. elegans. Cell* **1993**, *75*, 855–862.
3. He, L.; Hannon, G. J. MicroRNAs: Small RNAs with a Big Role in Gene Regulation. *Nat. Rev. Genet.* **2004**, *5*, 522–531.
4. Richard, W. C.; Erik, J. S. Origins and Mechanisms of miRNAs and siRNAs. *Cell* **2009**, *136*, 642–655.
5. Humphreys, D. T.; Westman, B. J.; Martin, D. I.; Preiss, T. Micro-RNAs Control Translation Initiation by Inhibiting Eukaryotic Initiation Factor 4E/cap and poly (A) Tail Function. *Proc. Natl Acad. Sci. USA* **2005**, *102*, 16961–16966.
6. Kiriakidou, M.; Tan, G. S.; Lamprinaki, S.; Planell-Saguer, M. D.; Nelson, P. T.; Mourelatos, Z. An mRNA m7G Cap Binding-like Motif Within Human Ago2 Represses Translation. *Cell* **2007**, *129*, 1141–1151.

7. Mathonnet, G.; Fabian, M. R.; Svitkin, Y. V., et al. MicroRNA Inhibition of Translation Initiation In Vitro by Targeting the Cap-binding Complex eIF4F. *Science* **2007,** *317,* 1764–1767.

8. Ding, X. C.; Grosshans, H. Repression of *C. elegans* MicroRNA Targets at the Initiation Level of Translation Requires GW182 Proteins. *EMBO J.* **2009,** *28,* 213–222.

9. Seggerson, K.; Tang, L.; Moss, E. G. Two Genetic Circuits Repress the *Caenorhabditis elegans* Heterochronic Gene Lin-28 After Translation Initiation. *Dev. Biol.* **2002,** *243,* 215–225.

10. Maroney, P. A.; Yu, Y.; Fisher, J.; Nilsen, T. W. Evidence that Micro-RNAs are Associated with Translating Messenger RNAs in Human Cells. *Nat. Struct. Mol. Biol.* **2006,** *13,* 1102–1107.

11. Petersen, C. P.; Bordeleau, M. E.; Pelletier, J.; Sharp, P. A. Short RNAs Repress Translation After Initiation in Mammalian Cells. *Mol. Cell.* **2006,** *21,* 533–542.

12. Thermann, R.; Hentze, M. W. Drosophila miR2 Induces Pseudo-polysomes and Inhibits Translation Initiation. *Nature* **2007,** *447,* 875–878.

13. Eulalio, A.; Huntzinger, E.; Izaurralde, E. GW182 Interaction with Argonaute is Essential for miRNA-mediated Translational Repression and mRNA Decay. *Nat. Struct. Mol. Biol.* **2008,** *15,* 346–353.

14. Behm-Ansmant, I.; Rehwinkel, J.; Doerks, T.; Stark, A., et al. mRNA Degradation by miRNAs and GW182 Requires Both CCR4:NOT Deadenylase and DCP1:DCP2 Decapping Complexes. *Genes Dev.* **2006,** *20,* 1885–1898.

15. Giraldez, A. J.; Mishima, Y.; Rihel, J., et al. Zebrafish MiR-430 Promotes Deadenylation and Clearance of Maternal mRNAs. *Science* **2006,** *312,* 75–79.

16. Wu, L.; Fan, J.; Belasco, J. G. MicroRNAs Direct Rapid Deadenylation of mRNA. *Proc. Natl Acad. Sci. USA* **2006,** *103,* 4034–4039.

17. Wakiyama, M.; Takimoto, K.; Ohara, O.; Yokoyama, S. Let-7microRNA-mediated mRNA Deadenylation and Translational Repression in a Mammalian Cell-free System. *Genes Dev.* **2007,** *21,* 1857–1862.

18. Yongli, B.; Christopher, S.; Yueh-Chin, C., et al. The CCR4 and CAF1 Proteins of the CCR4–NOT Complex are Physically and Functionally Separated from NOT2, NOT4, and NOT5. *Mol. Cell. Biol.* **1999,** *10,* 6642–6651.

19. Hai-Yan, L.; Vasudeo, B.; Deborah, C.A., et al. The NOT Proteins are Part of the CCR4 Transcriptional Complex and Affect Gene Expression Both Positively and Negatively. *EMBO J.* **1998,** *17,* 1096–1106.

20. Pillai, R. S.; Bhattacharyya, S. N.; Artus, C. G., et al. Inhibition of Translational Initiation by Let-7 microRNA in Human Cells. *Science* **2005,** *309,* 1573–1576.

21. Chendrimada, T. P.; Finn, K. J.; Ji, X., et al. MicroRNA Silencing Through RISC Recruitment of eIF6. *Nature* **2007,** *447,* 823–828.

22. Wang, Y.; Juranek, S.; Li, H.; Sheng, G.; Tuschl, T.; Patel, D. J. Structure of an Argonaute Silencing Complex with a Seed-containing Guide DNA and Target RNA Duplex. *Nature* **2008,** *456,* 921–926.

23. Valencia-Sanchez, M. A.; Liu, J.; Hannon, G. J.; Parker, R. Control of Translation and mRNA Degradation by miRNAs and siRNAs. *Genes Dev.* **2006,** *20,* 515–524.

24. Bagga, S.; Bracht, J.; Huntere, S., et al. Regulation by let-7 and lin-4 miRNAs Results in Target mRNA Degradation. *Cell* **2005,** *122,* 553–563.

25. Lim, L. P.; Lau, N. C.; Garrett-Engele, P., et al. Comparison of siRNA Induced Off-target RNA and Protein Effects. *RNA-Publ. RNA Soc*. **2007**, *13*, 385–395.

26. Brennecke, J.; Hipfner, D. R.; Stark, A.; Russell, R. B., et al. Comparison of siRNA Induced Off-target RNA and Protein Effects. *RNA-Publ. RNA Soc*. **2007**, *13*, 385–395.

27. Abrahante, J. E.; Daul, A. L.; Li, M., et al. The *Caenorhabditis elegans* Hunch-back-like Gene lin-57/hbl-1 Controls Developmental Time and is Regulated by microRNAs. *Dev. Cell* **2003**, *4*, 625–637.

28. Slack, F. J.; Basson, M.; Liu, Z.; Ambros, V., et al. The lin-41 RBCC Gene Acts in the *C. elegans* Heterochronic Pathway Between the let-7 Regulatory RNA and the LIN- 29 Transcription Factor. *Mol. Cell. Biol*. **2000**, *5*, 659–669.

29. Moss, E. G.; Lee, R. C.; Ambros, V. The Cold Shock Domain Protein LIN-28 Controls Developmental Timing in *C. elegans* and is Regulated by the lin-4 RNA. *Cell* **1997**, *88*, 637–646.

30. Johnston, R. J.; Hobert, O. A microRNA Controlling Left/right Neuronal Asymmetry in *Caenorhabditis elegans*. *Nature* **2003**, *426*, 845–849.

31. Zhao, Y.; Samal, E.; Srivastava, D. Serum Response Factor Regulates a Muscle Specific MicroRNA That Targets Hand2 During Cardiogenesis. *Nature* **2005**, *436*, 214–220.

32. Stark, A.; Brennecke, J.; Russell, R. B.; Cohen, S. M. Identification of Drosophila, MicroRNA Targets. *PLoS Biol*. **2003**, *1*, 1–13.

33. Lai, E. C.; Tam, B.; Rubin, G. M. Pervasive Regulation of Drosophila Notch Target Genes by GY-box, Brd-box, and Kbox- class MicroRNAs. *Genes Dev*. **2005**, *19*, 1067–1080.

34. Xu, P.; Vernooy, S. Y.; Guo, M.; Hay, B. A. The Drosophila microRNA mir-14 Suppresses Cell Death and is Required for Normal Fat Metabolism. *Curr. Biol*. **2003**, *13*, 790–795.

35. He, L.; Thomson, J. M.; Hemann, M. T., et al. A MicroRNA Polycistron as a Potential Human Oncogene. *Nature* **2005**, *435*, 828–833.

36. O'Donnell, K. A.; Wentzel, E. A.; Zeller, K. I., et al. c-Myc-regulated microRNAs Modulate E2F1 Expression. *Nature* **2005**, *435*, 839–843.

37. Lecellier, C. H.; Dunoyer, P.; Arar, K., et al. A Cellular microRNA Mediates Antiviral Defense in Human Cells. *Science* **2005**, *308*, 557–560.

38. Giraldez, A. J.; Cinalli, R. M.; Glasner, M. E.; Enright, A. J.; Thomson, J. M.; Basker-ville, S.;. Hammond, S. M.; Bartel, D. P.; Schier, A. F. MicroRNAs Regulate Brain Morphogenesis in Zebrafish. *Science* **2005**, *308*, 833–838.

39. Esau, C.; Kang, X.; Peralta, E., et al. MicroRNA-143 Regulates Adipocyte Differentiation. *J. Biol. Chem*. **2004**, *279*, 52361–52365.

40. Michael, M. Z.; O'Connor, S. M.; Van, H. P. N. G.; Young, G. P.; James, R. J. Reduced Accumulation of Specific microRNAs in Colorectal Neoplasia. *Mol. Cancer Res*. **2003**, 882–891.

41. Chang, T. C.; Yu, D.; Lee, S. Y., et al. Widespread MicroRNA Repression by Myc Contributes to Tumorigenesis. *Nat. Genet*. **2008**, *40*, 43–50.

42. Lin, S. L.; Chiang, A.; Chang, D.; Ying, S. Y. Loss of Mir-146a Function in Hormone Refractory Prostate Cancer. *RNA-Publ. RNA Soc*. **2008**, *14*, 417–424

43. Poy, M. N.; Eliasson, L.; Krutzfeldt, J., et al. A Pancreatic Islet-specific microRNA Regulates Insulin Secretion. *Nature* **2004,** *432,* 226–230.

44. Thai, T. H.; Calado, D. P.; Casola, S., et al. Regulation of the Germinal Center Response by microRNA-155. *Science* **2007,** *316,* 604–608.

45. Vigorito, E.; Perks, K. L.; Abreu-Goodger, C., et al. MicroRNA-155 Regulates the Generation of Immunoglobulin Class Switched Plasma Cells. *Immunity* **2007,** *27,* 847–859.

46. Rodriguez, A.; Vigorito, E.; Clare, S., et al. Requirement of Bic/microRNA-155 for Normal Immune Function. *Science* **2007,** *316,* 608–611.

47. Wienholds, E.; Kloosterman, W. P.; Miska, E., et al. MicroRNA Expression in Zebrafish Embryonic Development. *Science* **2005,** *309,* 310–311.

48. Yekta, S.; Shih, I. H.; Bartel, D. P. MicroRNA-directed Cleavage of HOXB8 mRNA. *Science* **2004,** *304,* 594–596.

49. Voorhoeve, P. M.; le Sage, C.; Schrier, M., et al. A Genetic Screen Implicates miRNA-372 and miRNA-373 as Oncogenes in Testicular Germ Cell Tumors. *Cell* **2006,** *124,* 1169–1181.

50. Fazi, F.; Rosa, A.; Fatica, A.; Gelmetti, V., et al. A Minicircuitry Comprised of MicroRNA-223 and Transcription Factors NFI-A and C/EBPα Regulates Human Granulopoiesis. *Cell* **2005,** *123,* 819–831.

51. Chang, S.; Johnston, R. J., Jr.; Frøkjær-Jensen, C. MicroRNAs Act Sequentially and Asymmetrically to Control Chemosensory Laterality in the Nematode. *Nature* **2004,** *430,* 785–798.

52. Sullivan, C. S.; Grundhoff, A. T.; Tevethia, S.; Pipas, J. M.; Ganem, D. SV40-encoded microRNAs Regulate Viral Gene Expression and Reduce Susceptibility to Cytotoxic T cells. *Nature* **2005,** *435,* 682–686.

53. Pandey, M.; Syed, S.; Dommez, I., et al. Coordinating DNA Replication by Means of Priming Loop and Differential Synthesis Rate. *Nature* **2009,** *462*(7275), 940–943. doi: 10.1038/nature08611. Epub 2009 Nov 18.

REVIEW QUESTIONS

1. Explain the different phases of mitosis and meiosis.
2. What is genetic code and what are its characteristics?
3. What is the significance of work done by Hargovind Khorana?
4. What do you understand by "central dogma of life."
5. What is needed to initiate translation in a prokaryotic and a eukaryotic cell?
6. What is the difference between DNA polymerases of prokaryotic and eukaryotic origin?
7. Explain rho dependent and rho independent termination of transcription.
8. What is regulatory RNA?
9. What is the difference between miRNA and siRNA?
10. What are the different proteins associated with regulatory RNAs?

11. What is codon bias and what is its role in evolutionary process?
12. Explain the different types of mutation.
13. How light is used to reverse the effect of mutation?
14. What are the different physical and chemical agents to induce mutation?
15. What is the role of DNA polymerase in mutation?
16. What are base analogs?
17. What are repair mechanisms present in a cell to safeguard against mutation?
18. What are nutritional mutants?
19. Explain suppressor mutation.
20. What is amber mutation?

CHAPTER 3

MOLECULAR BIOLOGY OF MICROORGANISMS

CONTENTS

ABSTRACT

In this chapter, we discuss the mechanism of propagation and survival of microorganisms, primarily bacteria, viruses, and yeasts. Bacteria are used regularly for molecular biological experiments, including cloning heterologous genes and growing plasmid DNA. We will also discuss the main mechanisms for exchanging DNA between individual bacterial cells. In addition to their chromosome, bacteria also may carry plasmids, the extrachromosomal circular DNA, which are frequently used. We will also discuss the CRISPR/Cas system, a prokaryotic acquired immunity that confers resistance to foreign genetic elements such as those in phages and plasmids, and is being exploited successfully for genome editing practices. Yeast is an ideal eukaryotic model because many processes occur in yeast cell can be readily studied leading to significant advances in understanding eukaryotic cellular processes.

Study of the molecular basis of the various physiological processes that occur in microorganisms has been a topic of extensive research. Bacteria were instrumental in the development of molecular biology as they are easy to culture and are small. Viruses can infect humans, plants and animals. Bacteriophages or "phage," have played important role in the development of the science of molecular biology and became "model organisms" for studying the biochemical processes of life.

3.1 BACTERIA

This section will discuss the bacterial gene transfer and their various molecular biological applications. Foreign genetic material can be introduced into a bacterial cell.

3.1.1 BACTERIAL CONJUGATION

Bacterial conjugation involves the transfer of genetic material between bacterial cells either by direct cell-to-cell contact or by forming a bridge-like connection between two cells.

Joshua Lederberg and Edward Tatum in 1946 studied two strains of *Escherichia coli* with different nutritional requirements.[1] They plated two strains A and B on the medium containing only unsupplemented minimal medium. Strain A could grow only if the medium was supplemented with methionine and biotin, and strain B could grow only if it was supplemented with threonine, leucine, and thiamine. Some of the dishes were plated only with bacterial strain A, some only with strain B, and some with a mixture of strain A and strain B bacteria that had been incubated together for several hours in a liquid medium containing all the supplements. No colonies were formed on plates containing either strain A or strain B alone, but the plates with both the strains produced significant number of colonies. This suggested that the genomes of the two strains had recombined to produce the prototrophs that could grow on unsupplemented medium.

The *E. coli* possess a circular DNA called the fertility factor or *sex factor* (F), that facilitates conjugation and gene transfer. Cells carrying the F plasmid are called F^+ cells, and those lacking it are F^- cells. Nearly 100 genes are present on the F plasmid which gives the plasmid several important properties:

1) The F plasmids are maintained even in a dividing cell population because they can replicate their own DNA.

2) The F^+ bacterial cells promote the synthesis of pili (singular, pilus) on their cell surface. *Pili* are minute proteinaceous tubules that allow the F^+ cells to attach to other cells and maintain contact with them; that is, to conjugate. The existence of specific sites both in the plasmid and in the chromosome helps in occasional homologous crossing-over. The plasmid gets integrated into the bacterial chromosome when a crossover between the two circular DNA molecules occurs. During integration, the entire host chromosome can be transferred into the recipient cell with the help of the F factor, along with its own integrated F DNA (Fig. 3.1).

The strains which have an F factor integrated into the chromosome of the host are called the *high frequency recombination* (*Hfr*) strains and differ from normal F^+ strains, which display a low frequency of recombination. The Hfr strains transfer their chromosomal markers efficiently and this attribute has been used for genetic mapping studies.

Sometimes the integrated F factor leaves the chromosome of an Hfr cell. This modified F is called *F'* cell, which carries a few host chromosomal genes along with it, and can transfer these specific host genes to a recipient (F^-) cell.

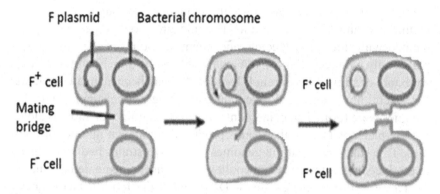

FIGURE 3.1 Bacterial conjugation of F⁺ and F⁻ cells. During conjugation, two bacteria proceed toward each other by the pilus. Next, a bridge forms between the two cells. One strand of plasmid DNA passes into the recipient bacterium and each single strand becomes double stranded again.

3.1.1.1 MECHANISM OF TRANSFER

All conjugations are by definition crosses of the donor Hfr and the recipient F⁻ cells. As the cells come together, the Hfr chromosome replicates and produces a single-stranded DNA molecule, which is then transferred linearly into the F⁻ cell. For any given Hfr strain, the order of transfer of markers are fixed and in a specific order. The replication and transfer begin at a specific point, called the *origin* and the genes closer to the origin are transferred first. The frequency of transfer of chromosomal markers from Hfr populations to F⁻ cells is higher than that of the F⁺ populations, because only a fraction of cells in an F⁺ population have F integrated into the chromosome.

It must be noted that in the cross between Hfr and F⁻, the F⁻ is not converted into Hfr or into F⁺, except in very rare cases, because the chromosome from Hfr breaks in nearly most of the cases before the F terminus is transferred to the F⁻ cell.

Once inside the recipient (F⁻ cells), the linear single-stranded DNA molecule acts as a template and is converted into a double-stranded DNA helix. This linear donor fragment is also called the exogenote,[2] and the chromosome inside the F⁻ cell is called the endogenote. Genetic exchange in prokaryotes takes place between one complete genome that of the F⁻, or the *endogenote* and an incomplete one from the donor, called the *exogenote*, thus creating a partial diploid, or merozygote[2–4] (Fig. 3.2).

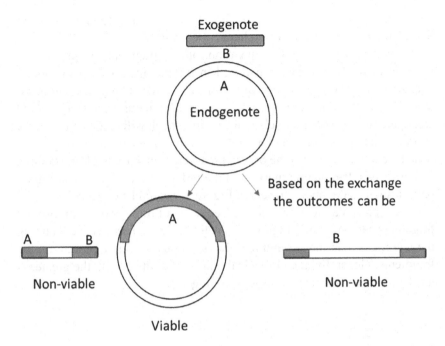

FIGURE 3.2 Crossover between exo- and endogenote occurs in a merozygote, (a). A partly diploid linear chromosome is formed after a single crossover, (b). A ring plus a linear fragment are formed as a result of an even number of crossovers, (c).

3.1.1.2 R FACTORS

The R factors are also called the resistance factors. Similar to the F factors in *E. coli*, the R factors are transferred rapidly through conjugation and confer resistance to the bacterial cells that harbor it. This ability of pathogenic bacteria was discovered in the 1950s in Japanese hospitals. Bacteria of the genus *Shigella* cause bacterial dysentery. Although this bacterium was initially sensitive to an array of antibiotics, those isolated from patients with dysentery were resistant to many antibiotics, including sulfanilamide, tetracycline, penicillin, chloramphenicol, and streptomycin. This phenotype was observed to be resistant to multiple drugs. The vector carrying these resistance factors proved to be a self-replicating element similar to the F factor.

The transfer of genetic information between individuals is inferred from the existence of recombinants produced from the cross. To obtain a recombinant that is stable, the transferred genes need to be integrated into the recipient's genome by an exchange mechanism.

However, at the end of conjugation, only a fragment of the chromosome from the donor gets transferred in the recipient, owing to spontaneous breakage of the mating pairs. The spontaneous breakage can occur at any time after transfer begins, which creates a natural transfer gradient and makes it less and less likely that a recipient cell will receive the genetic markers that appear later and are increasingly farther from the origin and hence are donated later in the order of markers transferred.[3,4] For example, in a cross of Hfr-donating markers in the order *met, ser, arg,* a distribution pattern such as the one represented Figure 3.3 could be expected.

The representative order and the frequency of transfer of chromosomal fragments can be seen in Figure 3.3. Note that more fragments containing the *met* locus than the *ser* and that the *arg* locus would be present on the fragments. Generally, the closer is the marker to the origin, the greater is its chance to be transferred through conjugation.

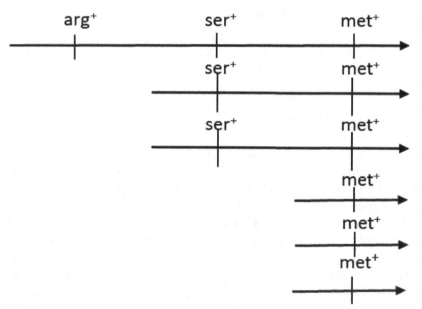

FIGURE 3.3 A schematic view of the transfer of markers over time. After bacterial mating, the frequency of recombinants for each metabolic marker as a function of time can be plotted. Transfer of the donor allele for each metabolic step depends on the time period for which conjugation is allowed.

3.1.1.3 DETERMINATION OF GENE ORDER BY GRADIENT TRANSFER METHOD

The order of transfer of genetic markers can be determined by the gradient transfer method, if an early marker is selected whose entry happens before the test markers that need to be sequentially ordered.[3,4] For example, let's take an Hfr strain that contains markers in the order *met, ser, leu,* and *his.* The cross of a *met⁺ ser⁺ leu⁺ his⁺ str*ˢ Hfr strain with *met⁻ ser⁻ leu⁻ his⁻ str*ʳ F⁻ strain, those recombinants growing on the minial medium without methionine but with amino acids, histidine and streptomycin, are selected. Those recombinants in the F⁻ strain are selected that are *met⁺* in the cross and in which the earliest marker that is transferred is *met* locus. The inheritance of other markers present in the Hfr can then be scored by testing on supplemental minimal medium that lacks one of the required nutrients. The result of the cross between the abovementioned strains could be:

$met^+ = 100\%$
$ser^+ = 70\%$
$leu^+ = 30\%$
$his^+ = 10\%$

The order of transfer gives a direct indication to the frequency of inheritance. For this method to work efficiently, it is important that it is used to screen for markers that enter after the selected marker (in this case, after *met*).

3.1.1.4 CASE STUDY

3.1.1.4.1 Interrupted Mating Experiments

The interrupted mating experiment technique with bacterial cell was worked out by François Jacob and Elie Wollman in the late 1957 while they were trying to demonstrate the mechanisms of gene transfer using *E. coli*. They crossed two *E. coli* strains whose genotypes are represented as below:

HfrH: *thr⁺ leu⁺ azi*ʳ *ton*ʳ *lac⁺ gal⁺ str*ˢ (prototrophic strain and sensitive to streptomycin).

F⁻: *thr⁻ leu⁻ azi*ˢ *ton*ˢ *lac⁻ gal⁻ str*ʳ (carries the gene for streptomycin resistance and a number of mutant genes).

The two strains were mixed in nutrient medium and incubated at 37°C to allow conjugation and a portion of the samples were analyzed at regular intervals after mixing, for further analysis. These samples were mixed vigorously for a few seconds to disrupt the mating cell pairs and then plated onto a medium. The minimal medium contained streptomycin to kill the Hfr donor cells. This procedure is called interrupted mating. The *str* cells then were checked for the presence of marker alleles that came from the donor. Those *str* cells bearing the test donor marker alleles must have received the alleles through conjugation, and are called exconjugants. A few interesting observations in this experiment are:

1) The donor alleles first appeared at a specific time interval in the F⁻ recipients after mating began.
2) The donor alleles appeared in a specific sequence.
3) The donor markers that entered later containing a specific donor allele gave rise to lesser number of progeny cells. After assimilating these results, Wollman and Jacob concluded that gene transfer occurs from a fixed point on the donor chromosome, termed the *origin (O)*, and continues in a linear fashion and that the farther the genetic markers from the origin, lesser is their chance of being transferred to the recipient.

The polarity of the Hfr chromosome is determined by the orientation in which F is inserted. The fertility factor thus exists in two states: (1) as a free cytoplasmic element F, or the plasmid state that is easily transferred to F⁻ recipients and (2) as a contiguous part of a circular chromosome, in an integrated state that is transmitted only very late in conjugation.

From this information, they constructed a tentative linear map based on different times of entry of each gene. Since only one F factor is integrated in each Hfr strain, this integration appears to occur at random, the Hfr strains vary with respect to the origin and direction of gene transfer. Therefore, by using different Hfr strains with more than one auxotrophic marker, in which the F factor has become integrated in different sites and in different orientations, it was possible to establish the complete genetic map of the *E. coli* chromosome.

The simplest way to arrange the genes of the *E. coli* chromosome is as a circular one. Moreover, the genetic distances between a particular pair of genes could be measured in time units, and were found to be constant (with marginal experimental error), independent of the Hfr strains used as donors. This map was determined to be 100 min long and provides information about the relative locations of the *E. coli* genes in the double-stranded circular chromosome. This is an example of successful execution of a complex experiment without the use of advance technology.

3.1.2 BACTERIAL TRANSFORMATION

Lateral gene transfer helps in the acquisition of genetic material and the studies on comparative analyzes of genomes show that it has been a major driving force in the evolution of prokaryotes.[1] Studies show that bacteria and archaea use the same mechanisms, that is, conjugation, transduction, and natural transformation, to acquire exogenous DNA. The transfer of DNA by natural transformation is initiated by the recipient cell.[5-7] When the cell gets genetically altered as a result of direct uptake and incorporation of exogenous genetic material from its surroundings through the cell membrane, it is said to be transformed. Transformation occurs naturally in some bacterial species. Transformation in bacteria was successfully demonstrated for first time in 1928 by British bacteriologist Frederick Griffith,[6] who discovered that a strain of *Streptococcus pneumoniae* could be made virulent after mixing with heat-killed virulent strains. He hypothesized that because of some "transforming principle" from the heat-killed strain could make the harmless strain virulent. This "transforming principle" was identified in 1944 as being genetic material by Oswald Avery, Colin MacLeod, and Maclyn McCarty.[7]

Most bacteria differ in their ability to uptake exogenous DNA. Those cells that can take up external DNA are called *competent*. Bacterial transformation may be natural or artificial. Bacteria can take up added DNA without any manipulation by natural transformation, for example, *Bacillus subtilis*. In artificial or engineered transformation, treatment of cells with chemical such as $CaCl_2$ or electric fields (electroporation) can induce competence, for example, *E. coli*.

Mechanism of transformation:

1) of the double-stranded donor DNA, one is degraded and the other enters the cell,
2) this single-stranded DNA then pairs with its homologous DNA in the chromosome leading to the formation of a triple strand,

3) a recombinant recipient may be formed upon crossover between this triple strand. If so, in the crossover region we can see donor and recipient DNA together-such DNA with different sequence on 2 strands is heteroduplex DNA, and

4) subsequent replication leads to equal number of daughter cells with the donor and recipient, that is, transformants and nontransformants, respectively (Fig. 3.4).

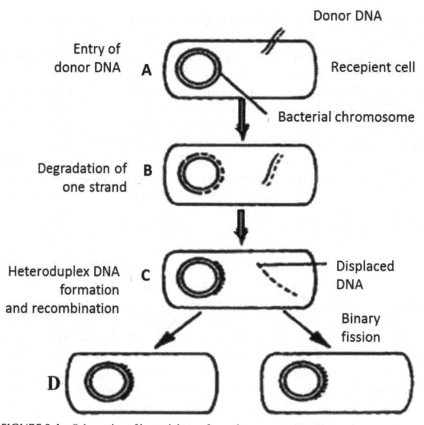

FIGURE 3.4   Schematics of bacterial transformation.

Transformation may be used to determine gene linkage and order using the principle of co-transformation. If two genes are located far away, they will be on different DNA fragments. The probability of simultaneous transformation is the product of the transformation probability of each gene. If the genes are close to each other (on the same DNA fragment) then the co-transformation frequency would be approximate to the frequency of a single gene.

3.1.2.1   REPLICATION AND MAINTENANCE OF BACTERIAL PLASMID

Replication is the process in which a genome is accurately copied. This process can be broken down into various steps that ensure genome duplication efficient. The most amazing fact of the entire process is that the machinery makes accurate copies the DNA, including proteins that are involved throughout the process.

Plasmids have been isolated from a wide range of prokaryotic organisms and two mechanisms of replication have been identified.

3.1.2.2   ROLLING CIRCLE MECHANISM

This mechanism is used by most small plasmids[8,9] (<12 kb). The process is initiated by a sequence-specific cleavage, at the nick site of the double-stranded origin (dso), of one of the parental DNA strands by an initiator Rep protein. This cleavage generates a 3'-OH end that allows host DNA polymerases to initiate the leading strand replication. This step hence, avoids the use of a RNA primer commonly used for initiation.

Elongation of the leading strand occurs by unwinding of the template by DNA helicase while the cleaved non-template strand is bound to single-stranded DNA binding (SSB) protein.[8–10]

Termination of one round of leading strand replication involves new cleavage event at the reconstituted nick site. This reaction is assumed to be catalyzed by the same Rep molecule that carried out the initiation cleavage and was attached to the parental strand parental strand during passage of the replication fork.

A trans-esterification then occurs that joins this 5' end to the 3' end generated in the termination cleavage, hence releasing the displaced parental strand as a circular single-stranded (ss) DNA. This replicative intermediate serves as the template for the synthesis of the lagging strand, which is initiated from a highly structured region of the ssDNA, called single-strand origin (sso).[9,10]

Thus, the outcome, in two separate steps is two circular dsDNAs containing either the newly synthesized leading or lagging strand and the complementary parental template strand (Fig. 3.5).

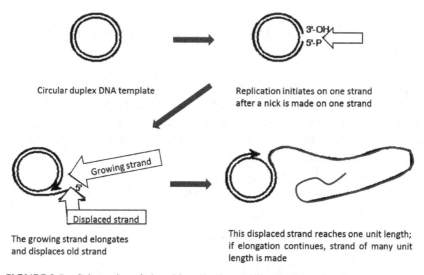

FIGURE 3.5   Schematics of plasmid replication: rolling circle mechanism.

DNA ligase and gyrase convert the daughter DNA into supercoiled forms.

Certain larger RCR plasmids contain additional genes and elements that contribute to their maintenance or help them transfer between host cells (Table 3.1).[8] The most ubiquitous is the MOB module, which is involved in the conjugative mobilization of the plasmid—it contains origin and mob or mobilizing proteins.

TABLE 3.1   A Few Characteristics Borne by the Genes on the Plasmids.

Characteristic	Plasmid examples
Fertility	F, R1, Col
Bacteriocin production	Col E1
Heavy-metal resistance	R6
Enterotoxin production	Ent
Metabolism of camphor	Cam
Tumorigenicity in plants	T1 (in *Agrobacterium tumefaciens*)

Some RCR plasmids also carry accessory genes encoding functions that can benefit the host conferring adaptation of the bacteria to their environment. Examples include antibiotic resistance and heat shock proteins.

3.1.2.3   THETA MECHANISM

Theta mechanism (TM) is employed by several larger plasmids. Here the lagging strand is synthesized discontinuously, and so intermediates have the shape of the Greek letter θ. Depending on the replicon, duplex melting can be either dependent on transcription or mediated by plasmid encoded *trans*-acting proteins (Reps).[10,11]

DNA replication through the TM involves melting of the parental strands, synthesis of a primer RNA (pRNA), and initiation of DNA synthesis by covalent extension of the pRNA. DNA synthesis is continuous on one of the strands (leading strand) and discontinuous on the other (lagging strand).

This mode of replication is similar to chromosomal replication in that leading and lagging strand are replicated coordinately, with discontinuous lagging strand synthesis. No DNA breaks are required for this mode of replication. For the elongation of plasmid DNA during replication, DNA Pol III is required, whereas for the early synthesis of the leading strand, DNA Pol I is involved.[10,11]

The termination of DNA replication is a regulated process with specific sequences called terminator site (ter) and prevents replication fork arrest mediated by the Tus protein.

3.1.2.4   STRAND DISPLACEMENT REPLICATION

This mechanism of replication has been studied in IncQ plasmids. Replication of occurs from two symmetrical and adjacent single-stranded origins (*ssiA* and *ssiB*) that are located on each DNA strand. Replication begins when these origins of replication are exposed as single strands. The melting of the DNA strand is dependent on two plasmid replication proteins, RepC and RepA, which recognize AT-rich regions preceding the *ssiA* and *ssiB* regions.[11] Another protein RepC recognizes directly repeated sequences of the origin adjacent to the AT-rich region, and RepA functions as a DNA helicase. Priming of DNA synthesis at these origins is catalyzed by RepB, the plasmid-specific primase.[11,12] The synthesis of each strand occurs continuously and results in the displacement of the complementary strand. Replication of this displaced strand is initiated at the exposed *ssi* origin. The use of RepA, RepB, and RepC plasmid proteins make the replication independent of the host.

3.1.3   TRANSDUCTION

Transduction was described for the first time by Norton Zinder and Joshua Lederberg in 1951.[13] It is the phenomenon by which bacterial DNA from one bacterium is introduced into another bacterial cell through a phage particle. Unlike transformation, transduction does not require physical contact between the cell that donates the DNA and the recipient cell.

3.1.3.1   GENERALIZED TRANSDUCTION

Any bacterial gene may be transferred to another bacterium via a bacterio-phage through the process of generalized transduction, and very rarely a small number of phages (1 out of 10,000 phages) carry the donor (bacterial genome) genome.[13,14] This is the packaging of bacterial DNA into a viral envelope. For this, the packaging may occur in two ways: recombination and headful packaging.

If upon entering a bacterium, the bacteriophage starts with the lytic cycle of infection, the virus takes control of the cell's machinery for use in replicating its own viral DNA. If the chromosomal DNA of the bacteria is inserted into the viral capsid which is used to encapsulate the viral DNA, this will lead to generalized transduction (Fig. 3.6).

The virus replication can result either in a nucleocapsid filled with genetic material, or viral packaging mechanisms may incorporate bacterial genetic material into the new virion. The new virus capsule which is now loaded with the portion of bacterial DNA continues to infect another host bacterial cell. This bacterial material may recombine into the genome of another bacterium upon infection.

When the new DNA enters this recipient cell it may meet with any one of three events may happen:

1)   The DNA is absorbed by the cell and be recycled.
2)   If the DNA was originally a plasmid, it will be re-circularized inside the new cell to become a plasmid.

3) If the new DNA has homology with any region of the recipient cell's chromosome, a process similar to the bacterial recombination may take place, and the DNA material is exchanged.

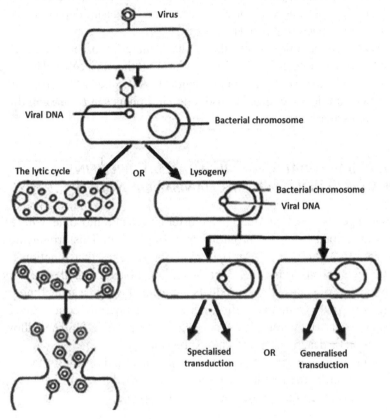

FIGURE 3.6 Schematic showing of the phage life specialized or generalized transduction. Bacteriophages utilize the host machinery for replication, transcription, and translation when they infect a host bacterial cell to make viral DNA or RNA and the protein coat and subsequently numerous virions.

3.1.3.2 SPECIALIZED TRANSDUCTION

Specialized transduction is the process by which a restricted set of bacterial genes is transferred to another bacterium.[13,15] It is the location of phage genome on the chromosome that determine the genes that get transferred (donor genes). Specialized transduction occurs when the excision of the prophage from the chromosome is imprecise so that bacterial genes lying adjacent to the prophage are also included in the excised DNA. The excised DNA is then packaged into a new virus particle, which then carries the DNA to a new bacterium, where the donor genes can be inserted into the recipient chromosome or remain in the cytoplasm, depending on the nature of the bacteriophage (Fig. 3.6).[15]

When the partially encapsulated phage material infects (is covalently bonded into the infected cell's chromosome) another host cell and becomes a "prophage," the partially coded prophage DNA is called a "heterogenote."[15]

For example, specialized transduction in λ phages in *E. coli* was discovered by Esther Lederberg.

3.1.4 BACTERIAL GENE REGULATORY PROTEINS: ORGANIZATION AND MECHANISM OF ACTION

Related genes are often found in a cluster on the chromosome, where they are transcribed from one promoter as a single unit. This arrangement is referred to as an operon. An operon contains genes that function in the same metabolism or process. Operons allow a bacterial cell to efficiently express sets of genes whose products are needed within a same time frame together. Apart from the coding genes, these operons containing DNA sequences called operators to which regulatory proteins bind and allow for activation or deactivation of gene expression.

Some regulatory proteins are *repressors* that bind operators and reduce transcription. Some regulatory proteins are *activators* that bind operators and increase the rate of transcription of the operon.

3.1.4.1 INDUCIBLE OPERON

The lac operon is an inducible operon that encodes enzymes for metabolism of the sugar lactose. It turns on only when the sugar lactose is present as the sole source of carbon. The operon has one promoter region and genes *lac* Z, *lac* Y, *lac* A, and *lac* I.[16] The *lac* Z, *lac* Y, and *lac* A produce β-galactosidase, lactose permease, and thiogalactoside transacetylase enzymes, respectively that are involved in lactose metabolism. The *lac* I gene produces a repressor protein that blocks RNA polymerase from binding to the promoter of the operon. If lactose is missing from the growth medium, the repressor binds very tightly to the operator and interferes with binding of RNAP to the promoter, hence repressing gene expression. In the presence of lactose as the sole carbon source, however, a lactose metabolite called allolactose, made from lactose by the product of the *lac* Z gene, binds to the repressor, causing an allosteric shift. Thus, when inactivated, the repressor cant bind to the operator; this allows RNAP to transcribe the *lac* genes and thereby leading to higher levels of the encoded proteins.

3.1.4.2   REPRESSIBLE OPERON

A second mechanism also involves cAMP-CAP.[17] The concentration of cAMP (cyclic adenosine mono phosphate) a signal molecule is inversely proportional to that of glucose. It binds to the CAP (catabolite activator protein), which in turn allows the CAP to bind to the CAP binding site—a 16 bp DNA sequence upstream of the promoter. This enables RNAP in binding to the DNA. In the condition when there is absence of glucose, the cAMP concentration is high and binding of CAP-cAMP to the DNA significantly enhances the β-galactosidase production.

The *trp* operon encodes enzymes for synthesis of the amino acid tryptophan (Trp) and is a repressible operon. This operon is expressed and repressed when high levels of Trp are present. *Trp* operon is also a cluster of genes controlled by a single promoter. This operon contains all the genes required for Trp synthesis. The operon consists of *trp* E, *trp* D, *trp* C, *trp* B, and *trp* A, which collectively code Trp synthetase; the enzyme which catalyses Trp biosynthesis.[16,17]

It is an example of repressible negative regulation of gene expression. In the upstream, the *trp*R gene expresses repressor for the *trp* operon; *trp*R gene is constitutively expressed at a low level. The repressor is synthesized as monomers that associate into dimers.

When Trp is present, these Trp repressor dimers bind to Trp, causing a change in the repressor conformation, allowing the repressor to bind to the operator. This action prevents binding of RNA polymerase to the promoter and hence prevents the expression of the operon.

This process is also aided by attenuation that responds to the concentration of charged tRNA. The *trp* operon contains the transcript (*trp*L), a leader sequence of at least 130 nucleotides. This transcript includes four short sequences which are numbered 1–4, each of which is partially complementary to the subsequent one. Thus, three distinct secondary structures called hairpins can form: 1–2, 2–3, or 3–4 (the last being a transcription terminator). Part of the leader transcript codes for a short polypeptide of 14 amino acids, termed the leader peptide that contains 2 Trp residues. If the ribosome attempts to translate this peptide while Trp levels in the cell are low, it will stall at either of the two trp codons causing the 2–3 hairpins that blocks 3–4 formation hence functions in antitermination.

If Trp levels in the cell are high, the ribosome will translate the entire leader peptide and stall at a stop codon blocking the 1–2 formation, permitting 3–4 formation that terminates expression.

3.1.5 INSERTION SEQUENCES IN BACTERIA

Insertion sequence (IS) elements are segments of bacterial DNA that can move from one position on a chromosome to a different position either on the same chromosome or on a different chromosome. The coding sequence is interrupted and the gene expression is inactivated when the IS elements appear in the middle of these genes. The genome of the standard wild-type *E. coli* is rich in IS elements. It is noteworthy that the IS can appear at any given locus under study, indicating that these elements are truly mobile, with a capability for transposition throughout the genome. A small set of ISs cause many different insertion mutations. The first sequence, the IS1 is 800-bp segment was identified in gal. The fact that sequences of more than 4000 different IS have been deposited in the specialized IS finder database indicates the number and diversity of known prokaryotic ISs since their discovery.[18] They have been observed to play important roles in the evolution of their host genomes and are involved in activating, sequestering, mutating, transmitting genes, and in the rearrangement of both plasmids and chromosomes.[18]

3.1.5.1 BACTERIAL TRANSPOSONS

Bacterial transposons are demarcated by inverted terminal repeats; when they insert into a DNA molecule, they create a duplication of sequences at the insertion site (a target site duplication).[19]

The bacterial chromosome and plasmids may contain several copies of an IS element. When a particular IS element is found on both a plasmid and a chromosome, homologous recombination may occur, a process that is found in conjugative R plasmids. These conjugative R plasmids have spread multiple drug resistance in bacterial populations, for example, the virulence plasmid of *Salmonella*, the F plasmid with *Tra* gene in *Staphylococcus*.

Composite transposons are bacterial cut-and-paste transposons, denoted by the symbol Tn and are created when two IS elements insert near each other, for example, Tn9, Tn5, and Tn10. Composite transposons consist of one or more genes for antibiotic resistance with two IS elements flanking this region.

The Tn3 elements are larger than the IS elements, such as composite transposons, contain genes that are not required for transposition (Fig. 3.7). These elements have simple inverted repeats at each end (not IS elements) and produce target site duplication when they transpose. The process of transposition of Tn3 is elaborated in Figure 3.8.

tnp A	tnp R	bla
Transposase	Resolvase/Repressor	B Lactamase

Inverted
Repeat   Inverted
                                                              Repeat

FIGURE 3.7 Genetic organization of Tn3, a replicative transposon. Tn3 fuses DNA molecules temporarily into a cointegrate; when the cointegrate is resolved, each of the constituent DNA molecules emerges with a copy of Tn3.

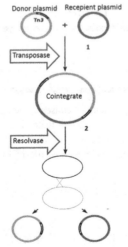

FIGURE 3.8 Transposition of Tn3. The Tn3 encoded transposase helps in cointegrate formation between donor and the recipient plasmids. Thus, each junction in the cointegrate contains a copy of the replicated Tn3. The resolvase expressed by *tnpR* mediates recombination between the two copied Tn3 elements and thus resolves the cointegrate. In the final Step, a copy of TN3 comes with the donor and recipient plasmid as they separate.

When an IS element inserts at a site, it causes target site duplication, as shown in Figure 3.9.

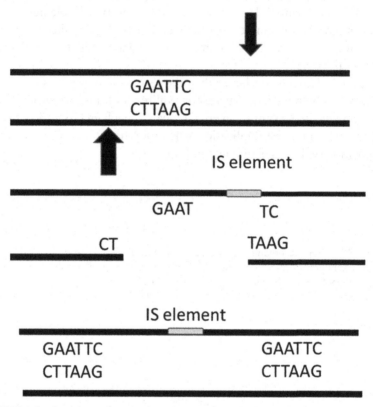

FIGURE 3.9   Insertion of an IS element causes target site duplication. Two strands of target DNA are cleaved at different sites, as indicated by the block arrows. The IS element is inserted in the gap that was created by the target DNA cleavage in step 1. The gaps are filled in by DNA synthesis on each side of the IS element. This produces an exact duplication of the target site.

3.2   VIRUSES

A virus is a small parasite that cannot reproduce by itself and can direct the host cell machinery to produce more viruses on infection.[20,21] In 1935, Wendell Stanley for the first time purified and partly crystallized tobacco mosaic virus (TMV); this was followed by crystallization of other plant viruses. The genetic material of most viruses is either single- or double-stranded DNA or RNA. The virus particle, called a virion, consists of the nucleic acid and an outer protein coat. The viruses contain only enough RNA or DNA to encode a range of proteins from 4–200 proteins.

3.2.1 PHAGES CONTAINING SMALL DNA

The best examples of small DNA phages are the filamentous M13 phages and the ΦX174. The viruses in this group are extremely simple and they do not encode most of the proteins required for their DNA replication, but depends on cellular proteins for this purpose. Thus, these viruses are particularly useful in studying the cellular proteins involved in DNA replication.

3.2.2 RNA PHAGES

Some phages infecting *E. coli* contain a genome composed of RNA. These phages are a good source of a pure species of mRNA because they are easy to grow in large amounts and their genomes also serve as their mRNA. These viruses, among the smallest known, encode only four proteins: two capsid proteins, an RNA polymerase for replication of the viral RNA, and an enzyme that dissolves the bacterial cell wall to allow release of the intracellular virus particles into the medium.

3.2.3 BACTERIOPHAGE

Bacteriophage consists of an outer protein coat/capsid that encapsulates an inner core of genetic material. Most bacterial phages show a lytic cycle where the host in which the virus replicates is lysed. For example, λ phage exhibits both lysis and lysogeny[22] (Fig. 3.7).

A key difference between the lytic and lysogenic phage cycles is that in the lytic phage, the viral DNA exists as a separate molecule within the bacterial cell, and its replication is not linked to that of the host bacterial DNA. The viral DNA is located within the host DNA in the lysogenic phage cycle, and, in both cases, the virus/phage replicates using the host DNA machinery (Figs. 3.7 and 3.10). However, in the lytic phage cycle, the phage is a free-floating molecule separated from the host DNA.

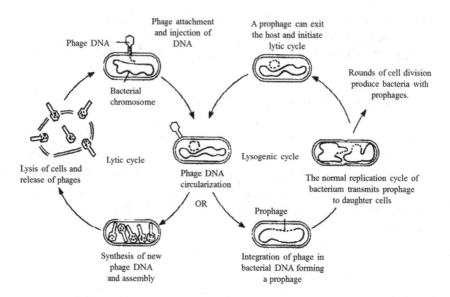

FIGURE 3.10   The life cycle of a phage. The bacteriophage undergoes either lytic or lysogenic cycle following infection. Most infected cells undergo lytic replication if the nutritional state of the host cell is favourable (left). Lysogeny is established (right), if the nutritional state of the host cell cannot support production of large numbers of phages. In the lysogenic phase, viral genes required for the lytic cycle are repressed, and host cell enzymes synthesize viral proteins that integrate the viral DNA into a specific sequence in the host cell chromosome where no host cell genes are disrupted. The prophage DNA replicates along with the host cell chromosome as the lysogen grows and divides. The progeny cells still maintain the repression of viral genes required for lytic replication. The prophage in a lysogen is induced at infrequent intervals, leading to expression of viral proteins that help in removing the prophage DNA from the host cell chromosome and derepressing the genes required for the lytic cycle. This results in resumption of normal lytic cycle.

A temperate bacteriophage carries out both lytic and lysogenic cycles. The phage replicates and lyses the host cell during the lytic cycle. However, the phage DNA is incorporated into the host genome, in the lysogenic cycle, where it is passed on to subsequent generations. Under certain conditions, the phage may undergo the lysogenic pathway by integrating its DNA into the host cell chromosome. In this state, the λ DNA is called a prophage and stays resident within the host's genome without harming the host. In the condition of presence of a prophage, the host is termed a lysogen, and when the lysogen enters a stress condition, this prophage may enter the lytic cycle.

3.2.3.1   LYTIC CYCLE

Replication of bacteriophages in a host bacterium shows the following steps (Fig. 3.9):

1) Attachment: Phages attach themselves on to a host bacterium's cell membrane using specific receptor molecules on the host that function for the firm attachment of the virus. Phages possess tail fibers on their base plate that help in the initial attachment.

2) Penetration: The tail fibers of the phage undergo contraction that brings the base plate and the host surface in close contact. The genetic material is then injected into the host cell upon the fiber contraction that leaves only the protein capsid outside the host. The empty protein capsids are termed as ghosts.

3) Synthesis: The injected viral DNA directs the replication of DNA and subsequent protein synthesis (capsid/base plate/tail fibers) using the replication and translation machinery of the host.

4) Assembly: The tail fibers are assembled with the base plate and then attached to the protein capsids. The replicated DNA is then assembled inside the capsids.

5) Release/Lysis: The newly assembled "virions" then are released by lysis of the host cell membrane/wall. This encompasses the use of an enzyme "endolysin" that helps in the perforation of the host cell wall. These newly released viruses then proceed to further lytic cycles of hosts.

3.2.3.2   LYTIC CYCLE WITHOUT LYSIS

Few viruses escape the host cell without bursting the cell membrane, but rather bud/extrude off from it by taking a portion of the membrane with them. M13 phages containing single-stranded DNA show this mechanism because it is otherwise a characteristic of the lytic cycle in other steps, although it is sometimes named the productive cycle. HIV, influenza, and other viruses that infect eukaryotic organisms generally use this method.

3.2.3.3   LYSIS OR LYSOGENY: A SOPHISTICATED GENETIC SWITCH

Lysogeny is characterized by integration of the phage nucleic acid into the host genome or formation of circular replicon within the host cytoplasm. In the lysogenic condition, the bacterium continues to live and reproduce normally.[22] The phage genetic material, also called the prophage, is transmitted to daughter cells at each subsequent host cell division, and are released at a later event (such as stress, i.e., UV radiation or the presence of certain chemicals), causing proliferation of new phages via the lytic cycle.[22,23]

3.2.3.4   THE GENETIC SWITCH IN PHAGES

The mechanism of this switch is discussed as follows:[22]

1) Transcription starts from the constitutive P_L, P_R and $P_{R'}$ promoters producing the "immediate early" transcripts. At first, these express the N and cro genes, producing N, involving N protein (Fig. 3.11).

2) Cro binds to $OR3$, preventing access to the P_{RM} promoter, preventing expression of the Ci gene. N binds to the two Nut (N utilization) sites, one in the N gene in the P_L reading frame, and one in the cro gene in the P_R reading frame.

3) The N protein is an antiterminator. Along with RNA polymerase and Nus proteins, this complex skips through most termination sequences. The extended transcripts (the "late early" transcripts) include the N and cro genes along with cII and $cIII$ genes. It must be noted that the λ-specific polymerase is not made; the $E.$ $coli$ RNA polymerase is used throughout the life cycle.

The stability of cII determines the lifestyle of the phage; stable cII leads to a lysogenic pathway, and the phage will go into the lytic pathway if cII is degraded. This is the life cycle that the phage follows after most infections, a condition when the cII protein due to degradation, is not present in a high enough concentration, and cannot activate its promoters.

1) The "late early" transcripts continue being written, including *xis*, *int*, *Q*, and genes for replication of the lambda genome (*OP*). *Cro* dominates the repressor site, repressing synthesis from the P_{RM} promoter (which is a promoter of the lysogenic cycle).
2) The *O* and *P* proteins initiate the phage chromosome replication.
3) Q another antiterminator, binds to *Qut* sites.
4) Transcription from the $P_{R'}$ promoter can now extend to produce mRNA for the lysis and the head and tail proteins.
5) Structural proteins and phage genomes self-assemble into new phage particles.
6) Products of the lysis genes *S*, *R*, *Rz*, and *Rz1* cause cell lysis. S is a holin,[24] a small membrane protein that, at a time determined by the sequence of the protein, suddenly makes holes in the membrane. *R* is an endolysin enzyme that escapes through the holes made by the holins and cleaves the cell wall. The membrane proteins *Rz* and *Rz1* form a complex that somehow destroys the outer membrane after the endolysin has degraded the cell wall.

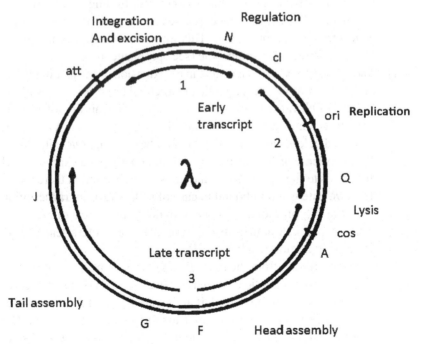

FIGURE 3.11   Lambda phage—genome organization.

3.2.3.5   RIGHTWARD TRANSCRIPTION

Rightward transcription expresses the O, P, and Q genes. O and P are responsible for initiating replication, and Q is another antiterminator that allows the expression of head, tail, and lysis genes from $P_{R'}$.

Q and N are similar to each other in their effects: Q binds to RNA polymerase in Qut sites and the resulting complex can ignore terminators. The lysogenic life cycle begins once there is a high enough concentration of cII protein activates its promoters, after a small number of infections.

1) The "late early" transcripts, including xis, int, Q, and genes for replication of the lambda genome continue being expressed.
2) The stabilized cII acts to promote transcription from the P_{RE}, P_I and P_{antiq} promoters.
3) The P_{antiq} promoter produces antisense mRNA to the Q gene message of the P_R promoter transcript, thereby switching off Q production. The P_{RE} promoter produces antisense mRNA to the cro section of the P_R promoter transcript, turning down cro production, and has a transcript of the cI gene. This is expressed, turning on cI repressor production. The P_I promoter expresses the int gene, resulting in high concentrations of int protein. This int protein integrates the phage DNA into the host chromosome (see "prophage integration").
4) The absence of Q protein results in no extension of the reading frame of the $P_{R'}$ promoter, thus no lytic or structural or proteins takes place. In case of elevated levels of int (much higher than that of xis) the lambda genome gets inserted into the hosts genome. When cI is produced, it leads to the binding of cI to the $OR1$ and $OR2$ sites in the P_R promoter, which turns off cro and other early gene expression. cI also binds to the P_L promoter, and turns off transcription there too.
5) The $OR3$ site is left unbound in case of lack of cro, so transcription from the P_{RM} promoter may occur, maintaining levels of cI.
6) Lack of transcription from the P_L and P_R promoters leads to no further production of cII and $cIII$.
7) The transcription from the P_{antiq}, P_{RE}, and P_I stop being promoted as cII and $cIII$ concentrations decrease, since they are no longer needed.
8) Only the promoters that are left active are P_{RM} and $P_{R'}$, the former produces cI protein and the latter a short inactive transcript. The genome remains inserted into the host genome in a dormant state (Fig. 3.12).

The lytic-lysogenic switch is a result of the proteins encoded by the viral genome. The switch is regulated by two regulatory proteins, the *cI* and *cro* regulators, as well as two promoters, *OL* and *OR cI* and *cro* define the lysogenic and lytic states, respectively, as a double stable genetic switch. *cI* maintains a stable lysogenic state, whereas *cro* activates the lytic cycle by indirectly lowering the levels of *cII* which activates *cI* transcription, hence blocking *cI* expression. The two regulatory proteins *cI* and *cro*, maintain this switch and the production of either determines the fate of the infected bacterium because an increase in cI protein promotes the lysogenic cycle, whereas increase in *cro* proteins promotes the lytic cycle. The regulation of the transcription of both the proteins is regulated by the *cI* protein itself (Fig. 3.12).

Gene	Protein	Site	Effect
cI	λ repressor that binds to $O_L + O_R$	$O_L + O_R$	λ repressor
N	Antiterminator that acts at $T_L + T_R$	$P_L + P_R$	Early and middle promoter
cII	Encodes activator of transcription from P_R and P_L	$t_L + t_R$	Lysogeny and transcription activator
cIII	*cIII*	P_Q	Regulatory protein *cIII*, lysogeny, and *cII* Stability
cro	Antagonist that acts at $O_R P_R P_M$ and $O_L P_L$ against repressor	P_I	*Int* promoter
Q	Q protein, activator of transcription of late genes	P_R	Antitermination for transcription of late lytic genes
Int	Integrase	*int*	Genome integration and excision
xis	Excision enzyme (*Xis*)	P_I, P_L	Genome excision

FIGURE 3.12 Lambda phage: the genetic switch. Different phage proteins, their functions and their site of action are mentioned in the table below the figure.

The action of the repressor protein is coupled with the attachment of the repressor dimer to the *OR*. *OR* is subdivided into three sites, *OR1*, *OR2*, and *OR3*, which are adjacent to each other, hence forming the right operator of the phage. The *cI* protein plays a role in both negative and positive control. The repressor binds to *OR2*, which turns off the *cro* gene. This prevents binding of the RNA polymerase to *PR*, the right promoter. The repressor partly covers the DNA vital for polymerase binding. Hence, with this binding of the repressor to the *OR2*, the RNA polymerase is unable to gain access to the recognition sequences for the promoter.

The lambda repressor also exhibits positive control, when although it still binds to the OR 2, it aids RNA polymerase binding and initiates transcription at PRM, which is the promoter regulating *cI* transcription. During negative control the repressor switches off its own genes, however in positive control, it does the opposite and only the phage genes are on, which increase transcription of its own genes.[23]

The binding of a *cI* dimer to *OR1* enhances binding of *OR2* to a second *cI* dimer, but not the affinity between *cI* and *OR3*. This leads to frequent occupying of the *OR1* and *OR2* by *cI*, in the presence of which only *cI* gene is transcribed. However, transcriptions of both genes are repressed at high concentration of *cI*.

When the host DNA is damaged (e.g., under UV irradiation), the *cI* protein is cleaved by certain protease promoted by the RecA protein; cleaved *cI* proteins cannot bind to the operators and progressing into the lytic cycle.

Cro plays an active role in switching lysogenic cells to the lytic state following induction. The protein binds noncooperatively to the three operator sites following its order of affinity, *OR3* > *OR2* = *OR1*. Thus, *Cro* ensures that the maintenance circuit for lysogeny does not come into play. Hence, following binding to the *OR3*, RNA polymerase binding to *PRM* is hindered and synthesis of repressor is inhibited.

3.2.4   CASE STUDY

GATEWAY cloning technology: This cloning technology[25] is based on the site-specific recombination system that is used by phage 1 to integrate its DNA, during lysogeny, in the *E. coli* chromosome. It was developed by Invitrogen, inc. The lysogeny is catalyzed by two enzymes: the phage 1 encoded integrase (Int) protein and the *E. coli* integration host factor (IHF) protein. Upon integration, the recombination between *att*B and *att*P sites generate *att*L and *att*R sites that flank the integrated phage 1 DNA. Two sequences were developed called *att*1 and *att*2 for each recombination site. These sites react very specifically with each other.

Two reactions constitute the GATEWAY cloning technology namely: The *LR reaction* is a recombination reaction between an entry clone and a destination vector, to create an expression clone, and the *BP reaction* is a recombination reaction between an expression clone (or an attB-flanked PCR product) and a donor vector to create an entry clone.

LR reaction: The LR reaction is used to create an expression clone. The recombination proteins cut to the *left* and *right* of the gene within the attL sites in the entry clone and ligate it to the corresponding attR site in the destination vector. The reaction is catalyzed by LR clonase enzyme mix contains recombination proteins *Int*, *Xis*, and the *E. coli*-encoded protein IHF (Fig. 3.13).

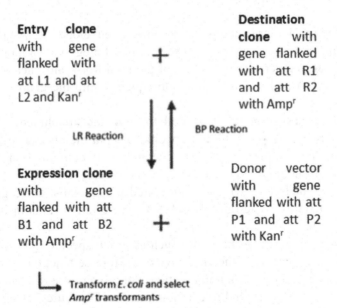

FIGURE 3.13 Schematic of gateway cloning technology. Donor vector: consists of counter-selectable genes flanked by *att*-P sites and it recombines with the gene of interest which have *att*-B sites on both sides to produce an entry clone. Entry clone: are formed by BP reaction and are used in LR reaction. It consists of the gene of interest and is used for producing expression clones. Expression clone: contains the gene of interest flanked by *att*-B sites. Destination clone: consists of counter-selection marker gene. It interacts with the gene from entry clone to give rise to expression vector by LR reaction.

BP reaction: The BP reaction is used to create an entry clone from expression clones or PCR products. Once a gene is flanked by attL sites (entry clone), it can be transferred into any number of "destination vectors" to generate new expression clones. The *BP clonase enzyme* mix facilitates mediates the BP Reaction and contains recombination protein *Int* and the *E. coli*-encoded protein IHF (Fig. 3.13).

Potentials: Gateway® cloning technique allows transfer of DNA fragments between different cloning vectors while maintaining the reading frame. It has effectively replaced the use of restriction endonucleases and ligases. Every subcloning reaction maintains the appropriate reading frame and is a fast process.[25]

3.2.5   AN OVERVIEW OF A FEW VIRUSES

3.2.5.1   RETROVIRUSES

These comprise a large and diverse family of enveloped RNA viruses.[26] The hallmark of this family is the replicative strategy which includes as essential steps reverse transcription of the viral RNA into linear double-stranded DNA and its integration subsequently into the genome of the cell.

Retroviruses enter the host cell through the attachment of their surface glycoproteins to highly specific plasma membrane receptors, following penetration, the RNA genome, still contained in a core complex of nonglycosylated proteins is reverse transcribed into a double-stranded DNA which is integrated into the host cell's genome (termed provirus). The retrovirus DNA is inserted at random into the host genome.

Certain proteins play key roles such as group specific (gag) proteins are major components of the viral envelope. Protease functions in proteolytic cleavages during virion maturation. Pol proteins are responsible for synthesis of viral DNA and integration into host DNA after infection and Env proteins play a role in association and entry of virions into the host cell.

The retroviruses are used as one of the preferred vectors for DNA technology because these viruses can integrate genetic material carried by the vector into recipient cells precisely, transduce a wide range of cell types from different animal species, and express the transduced genes at high levels.

3.2.5.2 LENTIVIRUS

This is a genus of viruses of the Retroviridae family (such as HIV, SIV) characterized by a long incubation period.[27] Lentiviruses can deliver a significant amount of viral RNA into the host cell and have the ability among retroviruses of being able to infect nondividing cells, so they are one of the most efficient methods of a gene delivery vector.[27,28]

The genome possesses three main genes coding for the viral proteins in the order: 5'-gag-pol-env-3'. Gag is a polyprotein and is an acronym for group antigens (ag), Pol is the reverse transcriptase, Env in the envelope protein. There are two regulatory genes, tat (involved in regulating viral replication) and rev.(that regulates protein expression). Additional products such as viral infectivity factor (Vif), is a protein found in HIV or viral protein R (Vpr) may be present.

Lentivirus is used to introduce a gene product into in vitro systems or animal models, and is thus used mainly as a research tool. The expression of short hairpin RNA (shRNA) reduces the expression of a specific gene finding applications in disease study and drug discovery. The introduction of a new gene into human or animal cells can be achieved by these viral vectors. This application finds use in gene therapy.

3.2.5.3 ADENOVIRUS

Adenoviruses are nonenveloped viruses containing double-stranded DNA as genetic material. The adenoviruses possess features that enable their use as viral vectors.[29,30] First, they are ubiquitous—isolated from a large number of different species, second, adenoviral vectors rapidly infect a broad range of human cells. Third, adenoviral vectors can accommodate relatively large segments of DNA and can transduce these transgenes in nonproliferating cells.

Adenoviral vectors allow for transmission of their genes to the host nucleus but do not integrate insert in chromosome. But, the adenoviral vector-based approach can be used in gene therapy treatment strategies in which only temporary protein expression is needed, for example, to elicit immune responses to pathological conditions. Hence, in conditions of chronic conditions, like cystic fibrosis, treatment will need to be repeated.

This finds potential in suicide gene therapy (also known as prodrug therapy) refers to the delivery of protoxins or chemotherapeutic agents into tumor cells.[31]

With the development of recently, "gutless" adenoviral vectors[32]—vectors that are devoid of all viral protein-coding DNA sequences there is further advancement. In this helper-dependent vector system, one vector (the helper) contains all the viral genes required for replication but in the packaging domain, it has a conditional gene defect. This defect makes it less likely that the DNA is packaged into a virion. The second vector contains only the ends of the viral genome, therapeutic gene sequences, and the normal packaging recognition signal, which allows this genome to be selectively packaged and released from cells. This system shows promise to offer further reduced toxicity and prolonged gene expression in animals.

3.2.6 PLAQUE ASSAY FOR VIRAL COUNTING

The number of infectious viral particles in a sample can be quantified by a plaque assay.[33] This assay is performed by culturing a dilute sample of viral particles on a media that is already covered with host cells. A local lesion, called plaque develops on the plate wherever a single virion initially infects a single cell. This is followed by counting the number of plaques that develop (Fig. 3.14). The virus replicates initially in this host cell finally lysing the cell, releasing many progeny virions that infect the neighboring cells on the plate. Enough cells are lysed after a few such cycles of infection, and a visible plaque is produced in the layer of remaining uninfected cells.

The plaque contains all the progeny virions that are derived from a single parent virus. While this type of plaque assay is in standard use for bacterial and animal viruses, plant viruses are assayed by counting local lesions on plant leaves inoculated with viruses. Our current understanding of molecular cellular processes has been greatly enhanced by analysis of viral mutants, which are commonly isolated by plaque assays.

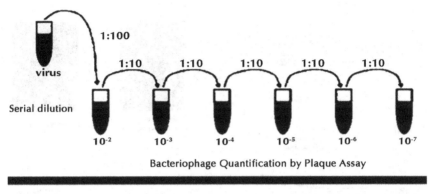

Bacteriophage Quantification by Plaque Assay

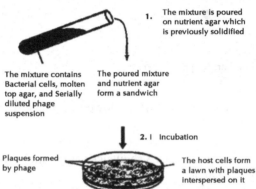

1. The mixture is poured on nutrient agar which is previously solidified

The mixture contains Bacterial cells, molten top agar, and Serially diluted phage suspension

The poured mixture and nutrient agar form a sandwich

2. | Incubation

Plaques formed by phage

The host cells form a lawn with plaques interspersed on it

FIGURE 3.14 Quantification of bacteriophage by plaque assay.

3.2.7 PHAGE DISPLAY TECHNIQUE

It is a selection method based on the presentation of peptides or proteins on the surface of bacteriophages. The DNA sequence encoding the peptide is expressed as a fusion protein with a gene coding for a surface protein of the phage, causing the phage to "display" the protein on its outside. These displaying phages are then screened against other peptides, proteins, or DNA sequences that enable the detection of interaction between the displayed protein and those other molecules (Fig. 3.15). Thus, the DNA sequences from random peptide libraries can be easily cloned into a phage vector and the selected peptides can be identified in very less time by sequencing of the DNA encoding the peptide inserts.

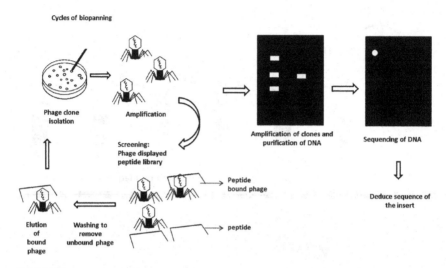

FIGURE 3.15   Flowchart for phage display.

The sequence of events that are followed in phage display screening are:

- The target proteins or DNA sequences are immobilized to the wells of a microtiter plate.
- The surface of the viral particle displays the genetic sequences in a bacteriophage library, which are expressed as fusions (of peptides or proteins or antibodies) with the bacteriophage coat protein.
- This phage-display library is added to the wells of the microtiter plate and the phage is allowed to bind for some time.
- The wells are then washed. Only the phage-displaying proteins that interact with the target molecules remain attached to the dish, while all others are washed away.
- This is followed by elution of the attached phage and its enrichment in suitable bacterial hosts.
- Steps 3–6 can be repeated to further enrich the phage library. This is followed by amplification of the DNA within in the interacting phage, sequencing to identify the interacting proteins or protein fragments (Fig. 3.15).

For example, the bacteriophage M13 is used extensively for phage display. The M13 bacteriophage is a filamentous phage of size 65 × 9300 Å, it has a circular single-stranded (ss) DNA genome of 6407 bases coated with 2700 copies of the major coat protein pVIII, 3–5 copies of pIII, and 5 each of copies of pVI, pVII, and pIX.[34,35] In the M13 life cycle:

- Infection begins with attachment of the pIII protein to the bacterial F-pilus. The pIII protein binds to the bacterial membrane protein after retraction of the F-pilus.
- Meanwhile, the phage DNA is transferred into the host cell and converted into a double-stranded DNA by host enzymes. The replicative dsDNA is amplified through a rolling circle mechanism (RCM) with the help of two phage proteins, pII and pX, leading to the accumulation of ssDNA forms.
- The pV protein covers the ssDNA, and prevents the conversion to dsDNA. At the membrane, the pI, pIV and Pxi make a complex and cover the DNA packaging structure leading to assembly of the phage particle and its release from the cell in a nonlytic manner.
- During the release, pV is exchanged for pVIII and the minor coat proteins are added.
- Phage displayed peptide libraries can be displayed on different coat proteins of the phage, but mostly pIII (5 copies) or pVIII (2700 copies) are used. Larger proteins are usually expressed through pIII. High affinity binders can be selected because of the low amount of pIII, whereas the high copy number of pVIII rather selects for low affinity binders (high avidity). The expressed proteins that interfere with infectivity can be expressed from a phagemid containing a minimal phage genome. This minimal phage genome would contain the bacterial and phage origins, an antibiotic resistance marker, and the phage gene used for display and a packaging signal. The phage proteins have to be provided in trans by a helper phage.

Some of the other commercially available phage libraries are: HIV-1 Env tailored phage libraries;[36] scFv phage libraries[37] help to obtain specific monoclonal antibodies without the need of immunization and generation of hybridomas.

3.2.8   CRISPR AND ITS ASSOCIATED PROTEINS: PROKARYOTIC ADAPTIVE IMMUNITY AGAINST VIRUSES

Clustered regularly interspaced short palindromic repeats (CRISPR) are segments of prokaryotic DNA containing short repetitions of base sequences which are followed by short segments of DNA, also called the "spacer DNA" obtained from previous exposures to a bacteriophage virus or plasmid. The CRISPR system from various species consists of several key related features such as:

(a) a CRISPR locus, which carries information about past infections and consists of short palindromic sequences of nearly 30 nucleotides that flank similarly sized spacer sequences that originate from invading nucleic acids[38] and

(b) four *cas* genes which are associated with CRISPR. These four *cas* genes bear homology to exonucleases and helicases.[38] These genes encode proteins that mediate CRISPR-based immunity and are usually associated with the CRISPR locus; in some cases, they are found elsewhere in the genome, the details of which are depicted in Figure 3.16.

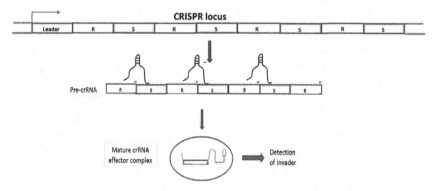

FIGURE 3.16   The CRISPR locus. The pre-crRNA that consists of repeat (R) and spacer (S) sequences is encoded by the CRISPR locus. Stem-loop structures represent a few folded repeat sequences. Spacer sequences are derived from invader DNA from previous exposures. Transcription of the CRISPR locus starts from the leader region (black arrow) generating pre-crRNA. This pre-crRNA is subsequently processed into crRNAs, and each being specific for one invader (Maier et al., 2012).

The CRISPR/Cas system is a prokaryotic acquired immunity that confers resistance to foreign genetic elements such as those in phages and plasmids. CRISPRs are found in approximately 40% of sequenced bacterial genomes and 90% of sequenced archaea.[41-43]

The CRISPR and CRISPR-associated (*Cas*) genes may function as a prokaryotic RNAi-like system that is essential in adaptive immunity, enabling the organisms to respond and eliminate invading genetic material.[39,40] For example, resistance against a bacteriophage can be acquired by *Streptococcus thermophilus* by integrating a fragment of the infectious viral genome into its CRISPR locus.[41] The *Streptococcus pyogenes* CRISPR-Cas system can be programed to cut any desired sequence, in vitro, by providing the Cas9 endonuclease and an appropriate guide RNA (gRNA).[43]

Therefore, by inserting sequences of their choice into CRISPR loci, researchers can use this bacterial immune system to put an end to the phage infections that could reduce productivity of commercially important strains.

The CRISPR-Cas system is classified into two classes:[43]

Class 1 system uses a complex of multiple Cas proteins that degrade foreign nucleic acids. Types I, III, and IV belong to this class.

Class 2 systems use a single large Cas protein to degrade foreign nucleic acids. Class 2 is divided into types II and V.

Most CRISPR-Cas systems have a Cas1 protein. The CRISPR-Cas systems may be compatible and share components because many organisms contain multiple CRISPR-Cas systems.[44,45]

The guide RNA consists of two RNAs, CRISPR RNA (crRNA) and trans-activating crRNA (tracrRNA). The tracrRNA can be combined to form a chimera, the single guide RNA (sgRNA), which is typically 100 nucleotides long. Twenty nucleotides at the 5' end base pair with complementary target DNA sequence by Watson–Crick base pairing. The Cas endonuclease is guided by this base pairing to cleave the target genomic DNA. The Cas9 recognition requires the remaining double-stranded structure at the 3' end (Fig. 3.17).

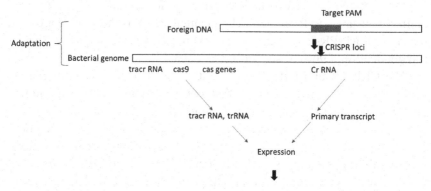

FIGURE 3.17 Bacterial adaptive immunity through Cas9. In the adaptation stage, foreign DNA is incorporated into the bacterial genome at the CRISPR loci. During biogenesis stage, the CRISPR loci is transcribed and processed into crRNA. For interference, Cas9 endonuclease makes a complex with a crRNA and separate tracrRNA cleaves foreign DNA containing a 20-nucleotide crRNA complementary sequence adjacent to the PAM sequence.

3.2.8.1   THE CRISPR-CAS9 SYSTEMS

The protospacer adjacent motif (PAM) is probably involved in the selection of protospacers in invading nucleic acid in type I and type II CRISPR-Cas systems.

The type II CRISPR-Cas system includes the CRISPR repeat-spacer array, a short guide RNA, and four or three Cas genes and is considered to be the "minimal CRISPR-Cas system". Synthetic sgRNA can be generated using in vitro transcription and can be delivered using a DNA vector expressing the sgRNA, or as synthetic RNA. The type II bacterial CRISPR-Cas9 system generates site-specific DNA breaks. DNA breaks are repaired by endogenous cellular mechanisms (Fig. 3.19).[46]

This mechanism can create (see Fig. 5.14, Chapter 5 for more information):

(a) Insertions or deletions are created at the site of the break,
(b) Nonhomologous end joining (NHEJ), which is mutagenic
(c) A change of a genomic sequence precisely through homologous recombination (HR).

Type III-B CRISPR-Cas system: The type III-B CRISPR-Cas system contains a crRNA and six Cas proteins (Cmr1 to Cmr6) that form an RNA silencing complex.

3.2.8.2 TARGET RECOGNITION AND DESTRUCTION BY CRISPRs

The three main stages of CRISPR Cas based immune defense are:[40,42-46]

Adaptation: During this stage, nucleic acid enters the cell and is immediately recognized as a foreign element. A piece of the foreign DNA, also called "the protospacer" gets selected and is integrated into the CRISPR locus. When the protospacer gets integrated into the CRISPR locus as part of the invading DNA sequence, it is called a spacer. Selection of a new spacer depends on the presence of a specific neighboring sequence, also known as the PAM. This type of adaptation is exhibited by CRISPR-Cas systems type I and type II. Briefly, the adaptation begins by recognizing the invading DNA by Cas1 and Cas2 and cleavage of a protospacer. The protospacer is then ligated to the direct repeat next to the leader sequence and single strand extension duplicates the direct repeat and repairs the CRISPR. The processing of crRNA and stages of interference occur differently in the three major CRISPR systems studied.

Expression: In this stage, the CRISPR locus is expressed and pre-crRNA is generated. This pre-crRNA is subsequently processed to short crRNAs, each being specific for a single foreign sequence. For this, the Cas cleaves the primary CRISPR transcript. In type I systems, Cas 6e/6f subunits cleave at the junction of ssRNA and dsRNA formed by hairpin looping in the direct repeat. Type II systems use a trans-activating (tracr) RNA to form dsRNA, which is further cleaved by Cas9 RNase III complex. The RNA within the repeats is cleaved by the Cas9 RNase III complex, and the spacers are cleaved at a fixed distance during maturation. A Cas6 homolog plays the role type III systems, and for cleavage, it does not require hairpin loops in the direct repeat. In type II and type III systems mature crRNAs are produced by the secondary trimming at either 5' or 3' end.[42-46]

Interference: The invading nucleic acid is cleaved. In type I systems the cascade complex is guided by the crRNA to the complementary DNA targets; the spacer sequence of the crRNA base pair with the foreign sequence from which it was derived. Thus, this defense is sequence specific. The invading DNA is thought to be cleaved by the Cas3 subunit,[40] and PAM is likely required for target recognition. In the type II systems, the Ca9 is thought to directly target the invading DNA after being loaded with crRNA, in a process that requires PAM. In type III systems, subtype

III A targets DNA and subtype III B targets RNA. When no base pairing to the 5' repeat fragment of the mature crRNA occurs, it results in no interference and vice versa.

3.2.9   ROLE OF VIRUSES IN GENE THERAPY

Many viruses can infect a large number of different cell types; therefore, they are often genetically modified to carry foreign DNA into a cell. This approach provides the basis for a growing list of experimental gene therapy treatments. Because of the extensive use of viruses in cell biology research and their potential as therapeutic agents, we describe the basic aspects of viral structure and function in this section.

For instance, the ability of virions to introduce their contents into the cytoplasm and nuclei of infected cells has been adapted for use in DNA cloning and offers possibilities in the treatment of certain diseases. The introduction of new genes into cells by packaging them into virion particles is called *viral gene transduction*, and the virions used for this purpose are called *viral vectors*.

Human adenovirus *recombinants* can be constructed in which potentially therapeutic genes replace the viral genes required for the lytic cycle of infection. The adenoviral vectors can introduce the engineered gene into the cells of tissues where they are applied because adenovirus can infect a very broad host range of human cells.[47] If a normal form of a protein is missing or defective causing a particular disease, the transduced gene can encode the normal protein. Such *gene therapy* may successfully treat the disease. However, the transduced gene is usually expressed only for a limited period (2–3 weeks). This significantly limits their usefulness in gene therapy.

Viruses that integrate their genomes into host cell chromosomes are also used to develop viral vectors. One of the main advantages of such vectors is that progeny of the initially infected cell also contain and express the transduced gene because it gets replicated and segregated in daughter cells along with the rest of the chromosome into which it is integrated. Retroviral vectors also are now widely used experimentally to generate cultured cells expressing desired proteins. However, a major concern with retroviral vectors is that they integrate randomly and might disturb normal expression of cellular genes.

Adeno-associated virus (AAV) is a "satellite" parvovirus that replicates only in cells that are coinfected with adenovirus or herpes simplex virus.[47] The ssDNA genome of AAV is copied into dsDNA by host cell DNA polymerase when it infects human cells in the absence of these "helper" viruses. It then gets integrated into a single region on chromosome 19, where it does not have any known deleterious effects. Studies are being conducted to adapt the AAV integration mechanism that operates in the absence of helper virus to the development of a safe and effective integrating viral vector.

3.3 YEAST

Yeast is a single-celled fungus which is classified on the basis of the cell, ascospores, colony, and physiological characteristics. A very well-known characteristic of yeast is the ability to ferment sugars to produce ethanol.

The most well-known and commercially significant yeasts belong to strains of *Saccharomyces cerevisiae*. It is a model organism for eukaryotic genetic models due to its non-pathogenicity, easy genetic manipulation, and knowledge of its genome sequence. This organism provides a highly suitable system to study basic biological processes that are relevant for many other higher eukaryotes. Strains of *S. cerevisiae* exhibit both a stable haploid and diploid state.[48] Recessive mutations can be conveniently isolated and expressed in haploid strains, and complementation tests can be studied in diploid strains.

S. cerevisiae contains haploid 16 well-characterized chromosomes; the total sequence of chromosomal DNA is 12,052 kb.[49]

Vegetative cell division of yeasts involves *budding*, or growth of a daughter cell as a "bud" and subsequent cell division separating the mother and the daughter cells. Haploid cells have buds adjacent to the previous one (radial budding), whereas diploid cells have buds at the opposite pole (axial budding).

Yeasts also show *sporulation*: Diploid cells produce 4 cells by meiosis that develops into endospores also termed as ascospores contained within a sac called ascus. (The majority of asci contain four haploid ascospores also hence, known as a spore tetrad).

3.3.1   YEAST CELL TYPES

There are two a and α mating types are under control of alleles.[48,50] The haploid vegetative cells undergo mitosis via budding. Fusion of the a and α mating types produce heterozygous pair of MATa/MATα. Subsequent sporulation produces ascus containing two MATa and two MATα haploid cells. Exposure to favorable conditions commences and mating of the MATa and MATα can occur.

The mating type locus MAT is located on chromosome III, which determines the mating type in yeast. The MATa locus encodes the transcriptional activator a1, and the MATα locus encodes the alpha1 activator and the alpha2 repressor.[48] The mating type locus also acts as a master regulator locus by controlling expression of many genes. Experimental crosses of MATa and MATα haploid strains are simply carried out by mixing equal amounts of each strain on a complete medium and incubating the mixture at optimum conditions. Prototrophic diploid colonies can be selected on suitable synthetic media if the complementing auxotrophic markers are present in haploid strains.

3.3.2   YEAST CAN SWITCH MATING TYPES

Two silent mating type loci on the same chromosome become activated when translocated to the MAT locus.[50] This translocation is a gene conversion initiated by the HO nuclease, that cuts similar to a restriction enzyme within the chromosomal active mating type locus. Laboratory yeast strains lack the HO nuclease and hence have stable haploid phases. Interestingly, only mother cells can switch. This ensures that after cell division two cells of opposite mating type are formed.

3.3.3   SPORE ANALYSIS AND GENETIC MAPPING

An important need of the yeast research community includes determination of the function of every gene in yeast. For this, several projects and approaches are being used. Microarray analysis for determination of simultaneous expression of all genes and the binding sites of transcription factors.[51] In addition, using yeast deletion analysis, more than 6000 deletion mutants are available for research.

Genetic markers are used to select diploids in genetic crosses, to follow chromosomes in genetic crosses, to select transformants when transformed with plasmids or integration of genes into the genome.

3.3.4 CROSSING STRAINS IN YEAST

The possibility to cross opposite mating type to a diploid strain and two haploid strains with different mutations makes the foundation of yeast genetics. Such genetic crosses were done to map genes on chromosomes. Genetic crosses are used to generate yeast strains with new combination of mutations, such as double, triple mutations.[52] Although the yeast genome is fully sequenced genetic screens are often performed to find genes/proteins that function in the same pathway than an already known gene/protein.

3.3.5 MAKING YEAST MUTANTS[53]

Mutations that alter the function of a certain protein are extremely useful in studying cellular systems. The phenotype of mutations can tell a lot about the function of a gene, protein or pathway. For this, random or targeted mutations can be carried out. While in *random mutagenesis* studies, usually the entire genome is targeted and the genes are linked to a certain function. This helps in finding new genes or new functions to known genes, a specific gene is altered or knocked out by a combination of in vitro and *in vivo* manipulation techniques in *targeted mutagenesis*.

Once mutants have been identified they are characterized and the genes are identified by involving the following steps: phenotypic analysis, establishing if a mutant is dominant or recessive, and by placing the mutants into complementation groups. Usually, one complementation group is equivalent to one gene.

3.3.6 YEAST TWO-HYBRID SYSTEMS

It is a technique to detect the interaction of two proteins in a yeast cell and can be used for selecting the interacting partner of a known protein.[54]

The method is very robust and is not restricted to yeast proteins only. The interacting partners can originate from any organism and the use of any yeast sequences is not mandatory.

A very important component of these two-hybrid systems are the transcription factors. These eukaryotic transcription factors have two separate domains, one for transcriptional activation and the other for DNA binding. While the two domains are present naturally on the same polypeptide chain, the transcription factor also functions if these two domains are brought in each other's vicinity by noncovalent protein–protein interactions. The gene fusions are constructed in such a way that the DNA-binding domain is linked to one protein, and the activation domain is linked to another protein. A reporter gene that is regulated by the transcription factor is expressed when interactions bring the DNA-binding and activation domains close together (Fig. 3.18).

Another two-hybrid system uses the DNA-binding domain from the E. coli lexA repressor protein and the lexA operator sequence.[55] In this system, the activator domain is a segment of an acidic peptide expressing E. coli DNA, which acts as a transcriptional activator in yeast when fused to a DNA-binding domain. The lexA transcriptional activator, such as the GAL4 system directs the protein into the nucleus through its nuclear localization signal. Yeast strains having lexA operators upstream of both yeast LEU2 gene and the E. coli lacZ have served as reporter genes.[56]

There are numerous possible applications to the two-hybrid system. Because of its sensitivity, relatively low-affinity interactions can be detected by yeast-two hybrid assays. The most common use is to directly detect interaction between two proteins. By mutagenesis and the use of a counter-selectable reporter, such as URA3, this system can be used for domains and residues characterization in two proteins that mediate interaction.[57] Also, when the two-hybrid system is used to screen with libraries of fused genes, these cloned genes encoding proteins that interact with the target protein can be obtained immediately. Yeast two-hybrid system can also be used to find proteins that regulate the interaction between two proteins.[58] Another important application is the use of system for drugs screening that inhibits the interaction between two proteins.[59]

A few other variations of yeast two-hybrid assay are:

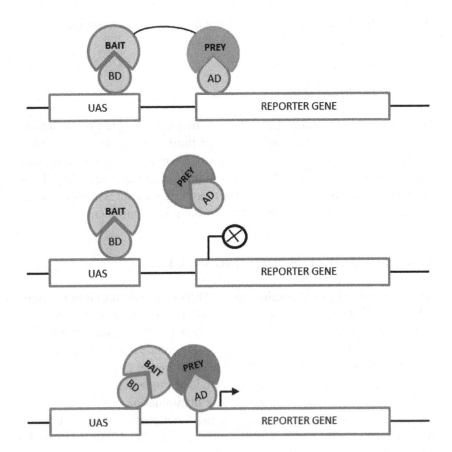

FIGURE 3.18 Principle of yeast-two-hybrid assay. The assay relies on the expression of a reporter gene (such as lacZ), which becomes active by the binding of a particular transcription factor. The transcription factor consists of a DNA-binding domain (BD) and an activation domain (AD). The protein of interest (query) fused with BD is known as the bait, and the protein library fused with the AD is referred to as the prey. To observe a positive reporter gene expression, a transcriptional unit must be present at the gene locus. This is possible only if bait and prey interact.

3.3.6.1   THE YEAST 1-HYBRID ASSAY

This assay is used for studying protein–DNA interactions. A query protein directly fused with the AD domain is expressed in yeast strains possessing several target DNA sequences upstream of the reporter gene. The AD domain will activate reporter gene if the query binds to a particular target sequence.

3.3.6.2   YEAST 3-HYBRID

These assays study interactions that are mediated by the protein and an RNA molecule or some other protein (a third component). In this case, there is an indirect interaction between bait and prey. For example, they bind to an RNA molecule, but with different sequence specificity. Thus, bait and prey would interact and drive reporter gene expression only in the presence of a specific RNA molecule, or the third component.

3.3.7   YEAST AS A MODEL ORGANISM

Perhaps most principal cellular systems function in a similar way across eukaryota, that is, in yeasts and human. The yeasts *S. cerevisiae* and *Schizosaccharomyces pombe* are regarded as model organisms in molecular biology.[60] The complete sequence of its genome is used as a reference toward the human sequences and that of other higher eukaryotes. Furthermore, genetic analysis in yeasts has provided fundamental insight in cell cycle control. The ease of yeast genetic manipulation enables its convenient use in analyzing and dissecting the functionality of gene products from other eukaryotes. Yeast is also used as a model organism to study: Signal transduction, morphology switching, vesicular transport, proteasome studies, aging, and functional genomics.

A well-defined genetic system, a highly versatile DNA transformation system, rapid growth, dispersed cells, the ease of replica plating and mutant isolation are a few properties that make yeast particularly suitable for biological studies. The yeast *S. cerevisiae*, unlike many other microorganisms, is viable with numerous markers and can be handled with little precautions because it is nonpathogenic.

Moreover, the development of DNA transformation has made yeast particularly accessible to gene cloning and other genetic engineering manipulations. Complementation from plasmid libraries can identify any structural genes corresponding to any genetic trait. Plasmids can be introduced into yeast cells either as replicating molecules or by integration into the genome via homologous recombination. Exogenous DNA with even minimal homologous segments can thus be directed at will to specific locations in the genome.[61]

3.3.8 YEAST BIOTECHNOLOGY

The largest biotechnology businesses worldwide employ the yeast fermentation technology for brewing, baking, winemaking, and industrial alcohol production. The industrial yeast strains are diploid, polyploid, or even aneuploid, and many appear to be cross-species hybrids are usually difficult to work with. Although various possible improvements to the fermentation processes can be made, the biology of yeast becomes the limiting factor. Hence, there are many attempts to improve yeasts-like ability to degrade polysaccharides, ability to hydrolyze saccharides, better osmotic and alcohol tolerance; better productivity and less byproducts during starvation, ability to kill competing bacteria and yeasts (cleaner fermentation and wine taste), ability to degrade different sugars at once through diminished catabolite repression (better leavening) and freeze tolerance after initiation of fermentation, etc.

3.3.8.1 HETEROLOGOUS EXPRESSION IN YEAST

The production of proteins is required for research, such as for purification and structural analysis, industry, the production of enzymes for the food and paper industry or for research and diagnostics, the pharmaceutical industry for the production of vaccines. Various different expression hosts, such as bacteria and yeasts are used for protein production., the yeast *S. cerevisiae* and *Pichia pastoris* have some attractive features over *E. coli* is still the primary choice for production of heterologous proteins.[62] Proteins produced in yeast, unlike those produced in *E. coli*, lack endotoxins and have several posttranslational processing mechanisms that allow the expression of several human or human pathogen-associated proteins with appropriate authentic modifications. Such posttranslational modifications include proteolytic processing, particle assembly, amino-terminal acetylation, and myristoylation. In addition, heterologous proteins secreted from engineered yeast strains are properly cleaved and folded and are amenable to easy harvest from the culture media. The use of either homologous or heterologous signal peptides allows proper maturation of secreted products by the host yeast machinery.

High-level expression of foreign protein is not necessarily required from the cloned cDNAs from other organisms and the study of their function using yeast as a surrogate. Even physiological quantities of the protein are sufficient in a correctly modified form and localized in the cell such that their activity reflects the activity in the parent organism.

Commercial and laboratory preparations of proteins generally require expression vectors that produce high amount of protein. There are various expression vectors currently available for producing heterologous proteins in yeast, and these are derivatives of the YIp, Yep, and YCp plasmids.[63] For the expression, the cDNA, synthetic DNA or intron-free genomic DNA lacking are inserted in a suitable vector. There is no ribosome binding site in any *S. cerevisiae* mRNA species studied so far. Promoters used in these expression vectors include a transcription initiation site and variable amounts of DNA encoding the 5′ untranslated region. Most of the yeast expression vectors do not contain an ATG in the transcribed region of the promoter, so, the heterologous gene must contain an ATG that establishes the correct reading frame and correspond to the first AUG of the mRNA. This is because translation almost always initiates at the first AUG on mRNAs of yeast and that from other eukaryotes.[64] Moreover, most commercial yeasts have a marked degree of preferential codon usage and the heterologous gene is likely to be expressed better if it has the same bias.

Numerous normal and altered *yeast promoters* are used, which are chosen because of their high activity and their regulatory properties. A few of these promoters have been derived from genes encoding alcohol dehydrogenase I,[65] enolase, glyceraldehyde-3-phosphate dehydrogenase, phosphoglycerate kinase, triose phosphate isomerase, galacokinase, repressible acid phosphatase,[66] a mating factor,[67] etc. and have been used depending on the specific heterologous gene.

The species *P. pastoris* is one of the most productive known yeast. It catabolizes methanol and uses the promoter for methanol oxidase which is extremely strong and is methanol induced. In *S. cerevisiae,* usually, the promoters used are of genes encoding glycolytic enzymes such as *PGK1* and *TPI1* or a regulated promoter such as that of *GAL.*[66,67]

3.3.8.2   CLONING IN YEAST

The first ever successful transformation of yeast with a foreign DNA in 1978 was the beginning of the era of yeast molecular genetics. However, there are several problems associated with yeast cloning. All the transformation protocols are much less efficient as *E. coli* transformation. Although replicating plasmids can be maintained by yeast, their copy number is much less than in *E. coli* (usually 1–50/ cell). Another complicating situation arises while gene cloning from a library because yeast can maintain more than one type of plasmid at the same time, it can also be very useful to transform yeast with two different plasmids simultaneously, for instance for a method called plasmid shuffling.[68]

Plasmid preparation from yeast is very ineffective. Therefore, *E. coli* is used as a plasmid production system for cloning in yeast. These plasmids are constructed in vitro and transformed into *E. coli*. The constructions are confirmed, in a way which is similar to that in bacteria. The confirmed plasmids are overproduced in bacteria and then transformed into yeast. The plasmids used in this process are called "shuttle vectors."[69]

3.3.8.3   YEAST VECTORS

A wide range of vectors are available to meet various requirements for insertion, deletion, alteration, and expression of genes in yeast. Most vectors used for yeast studies are shuttle vectors, which contain sequences permitting them to be amplified, altered in vitro, and subsequently selected and propagated in *E. coli*. The most common yeast vectors originate from pBR322 and contain its origin of replication (ori), promoting their maintenance in *E. coli in* high copy numbers, and have selectable antibiotic markers such as the b-lactamase gene, *bla* (or *Amp*R), and sometime to tetracycline-resistance gene, tet or (TetR), conferring resistance to ampicillin and tetracycline, respectively.

Homologous recombination system in yeast has a very efficient and can be used for cloning. Some of the plasmids used for this purpose are as follows:

Yeast integrative plasmids (YIp): YIp have the backbone of a *E. coli* vector such as pBR322, pUC19, pBluescript, and a yeast selection marker such as *URA3*, *HIS3*, *TRP1*, and *LEU2*. These vectors lack any yeast replication origin and are propagated only through integration into the genome by homologous recombination.[63,70] The site of integration can be targeted by cutting the yeast segment in the YIp with a restriction endonuclease. The yeast strain is then transformed with the linearized plasmid. The linear ends are recombinogenic and direct integration to the site in the genome that is homologous to these ends. Integration results in the duplication of the target sequence which flanks the vector. Characteristically, the YIp vectors integrate as a single copy. However, multiple integration events occurs at low frequencies, a property that can be exploited to construct stable strains that can overexpress specific genes. The integrated plasmids propagate stably but occasional pop out by recombination between the duplicated sequences is also observed. YIp is used for integration only.

Yeast episomal plasmids (YEp): The YEp vectors replicate autonomously because of the presence of either a full copy of the 2 μm plasmid or an ori (2 μm ori) on a segment of the yeast 2 μm plasmid.[63] The 2 μm ori is responsible for the high transformation frequency and high copy number of YEp vectors. YEp plasmid vectors are frequently used for overproducing gene products in yeast.

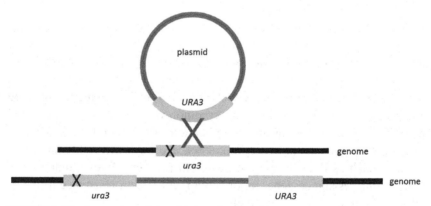

FIGURE 3.19   Integration of plasmids into the yeast genome.

Yeast centromeric plasmids (YCp): The YCp also consists of the backbone of a *E. coli* vector such as pBR322, pUC19, and pBLUESCRIPT and a yeast selection marker such as *URA3*, *HIS3*, *TRP1*, and *LEU2*. The YCp has a chromosomal replication origin for yeast, ARS (for autonomously replicating sequence) and have the centromere of a yeast chromosome.[63,65–67] The stability and low copy number of YCp vectors make them the ideal choice for cloning vectors, for construction of yeast genomic DNA libraries, and for investigating the function of genes altered *in vivo*. Hence, they are propagated stably at low copy number, typically one per cell. Thus, YCps are used for low copy expression and YEp are used for overexpression.

Yeast artificial chromosomes (YAC): YAC cloning systems are based on yeast linear plasmids, denoted YLp, containing homologous or heterologous DNA sequences that function as telomeres (TEL) *in vivo*, as well as contain the yeast ARS and CEN segments, which are origins of replication and centromeres, respectively.[63,65–67] Manipulating YLp linear plasmids in vitro is complicated by their inability to be propagated in *E. coli*. However, circular YAC vectors have been specially developed for amplification in *E. coli* (Fig. 5.8, Chapter 5). One common type of YAC vector can be propagated in *E. coli*, as it contains telomeric sequences in an inverted orientation. After amplification in *E. coli* the plasmid is digested and linearized with a restriction endonuclease before transforming yeast. Yeast are transformed by this linear structure at high frequencies, although the transformants are unstable.

3.3.8.4 *EXTRANUCLEAR GENOMICS OF YEAST*

Yeast possess 4–5 mitochondrial DNA (80 kb) molecules in a nucleoid, with each nucleoid having 10–30 such nucleoids.[71] Mitochondrial DNA encodes components of its translational machinery and nearly 15% of the mitochondrial proteins. The information coded by the DNA includes tRNAs, rRNAs, and subunits of enzymes such as ATPase, cytochrome oxidase, etc. The mitochondrial proteins, however, are encoded by nuclear genes, synthesized on cytoplasmic ribosomes and then transported into the mitochondria.

Boris Ephrussi in 1940s discovered few yeast cells of smaller size than the wild-type colonies on solid medium, they were called "petite" colonies (petite—small) and the wild-type "grandes" (big in French). These cells were found to lack mitochondrial functions due to changes in mitochondrial DNA that utilize fermentation for energy and not aerobic respiration and hence grow slowly.[72]

Types of petites:

Nuclear petites: There is genetic alteration in nuclear DNA that causes abnormal subunits of several mitochondrial proteins. When these are crossed with wild-type grande colonies we see 2:2 seggregation of petite:grande in the ascus (characteristic Mendelian genetics).

Neutral petites: Approximately 99–100% of mitochondrial DNA is missing and they are unable to perform aerobic respiration. They are designated as rho−N.

When these, that is, rho−N are crossed with wild-type rho+N, the diploid rho−N/rho+N produce grande colonies. Upon meiosis, the tetrads show 0:4 petite:grande. This is classical uniparental inheritance as the progeny shows only phenotype of one parent.

Suppressive petites: They begin as mitochondrial DNA deletions, due to a corrective mechanism undeleted sequences duplicate to restore the amount of DNA. Due to these genetic rearrangements, enzymes are deficient leading to lack of aerobic respiration and petite colonies. These petites are designated as rho-S. When these, that is, rho-S are crossed with wild-type rho+S, the diploid rho-S/ rho+S produce colonies with intermediate respiration between petite and grande. Subsequent meiosis and sporulation lead to 4:0 ratio of petite:grande while the nuclear genes segregate 2:2.

3.4   SUMMARY

Bacteria are one of the ideal organisms for research because of their rapid division time, and easy maintenance. *E. coli* is used regularly for cloning heterologous genes and growing plasmid DNA. Transformation, transduction, and conjugation are the main mechanisms for exchanging DNA between individual bacterial cells. Bacteria may carry smaller extrachromosomal circular DNA, called plasmids, in addition to the circular chromosome. Plasmids are frequently used in all fields of research as vectors for cloning and amplifying DNA from many different organisms.

Bacteriophages are viruses that infect the bacteria and undergo lytic or lysogenic cycle. The phage particles can be isolated and large quantities of the cloned DNA can be recovered experimental uses. The CRISPR/Cas system is a prokaryotic acquired immunity that confers resistance to foreign genetic elements such as those in phages and plasmids. For example, bacteria can acquire resistance against a bacteriophage by integrating a fragment of the infectious viral genome into its CRISPR locus. This mechanism is exploited successfully for genome editing practices.

Yeast is an ideal organism model because similar to eukaryotic cells, many processes occur in yeast and can be more readily studied. Yeasts generally exists two different mating types, "a" and "α" in *S. cerevisiae*. Significant advances in understanding eukaryotic cellular processes have been made as a result of the yeast cell's ability to introduce and select for specific mutations. Moreover, yeast cells can be easily transformed with plasmids carrying heterologous genes expressing proteins of interest.

3.5 QUESTIONS

1) Write short notes on:

(a) Bacterial transformation and explain exo- and endogenotes
(b) Host range of viruses
(c) Virion-associated enzymes
(d) The genetic switch of phages
(e) Applications of CRISPR
(f) Gateway technology
(g) Viruses used for gene therapy

2) Differentiate between rolling circle and theta replication of plasmids.
3) What characteristics make yeast a model organism?
4) How does a bacterium acquire immunity against foreign genetic elements? (Hint: this technology is now used for genome editing).

KEYWORDS

- **bacteriophages**
- **translation**
- **vectors**
- **regulatory proteins**
- **recombination**

- yeast
- CRISPER
- Cas

REFERENCES

1. Tatum, E. L.; Lederberg, J. Gene Recombination in the Bacterium *Escherichia coli*. *J. Bacteriol.* **1947**, *53*, 673–684.
2. Komeda, Y.; Silverman, M.; Simon, M. Genetic Analysis of *Escherichia coli* K-12 Region I Flagellar Mutants. *J. Bacteriol.* **1977**, *131*(3), 801–808.
3. Griffiths, A. J. F.; Miller, J. H.; Suzuki, D. T., et al. *An Introduction to Genetic Analysis, 7th ed.*; W. H. Freeman: New York, 2000.
4. Jones, R. T.; Curtiss III, R. Genetic Exchange Between *Escherichia coli* Strains in the Mouse Intestine. *J. Bacteriol.* **1970**, *103*(1),71–80.
5. Johnsborg, O.; Eldholm, V.; Håvarstein, L. S. Natural Genetic Transformation: Prevalence, Mechanisms and Function Research. *Microbiology* **2007**, *158*, 767–778.
6. Griffith, F. The Significance of Pneumococcal Types. *J. Hyg-Cambridge.* **1928**, *27*(2), 113–159 (January 1928).
7. Avery, O. T.; MacLeod, C. M.; McCarty, M. Studies on the Chemical Nature of the Substance Inducing Transformation of Pneumococcal Types: Induction of Transformation by a Desoxyribonucleic Acid Fraction Isolated from Pneumococcus Type III. *J. Exp. Med.* **1944**, *79*(2), 137–158.
8. Khan, S. A. Rolling-circle Replication of Bacterial Plasmids. *Microbiol. Mol. Biol. Rev.* **1997**, *61*(4), 442–455.
9. del Solar, G.; Giraldo, R.; Ruiz-Echevarría, M. J.; Espinosa, M.; Díaz-Orejas, R. Replication and Control of Circular Bacterial Plasmids. *Microbiol. Mol. Biol. Rev.* **1998** June, *62*(2), 434–464.
10. O'Donnell, M.; Langston, L.; Stillman, B. Principles and Concepts of DNA Replication in Bacteria, Archaea, and Eukarya. *Cold Spring Harb. Perspect. Biol.* **2013**, *5*, a010108.
11. Lilly, J.; Camps, M. Mechanisms of Theta Plasmid Replication. *Microbiol. Spectr.* **2015**, Feb, *3*(1), PLAS-0029-2014.
12. Manosas, M.; Spiering, M. M.; Ding, F.; Bensimon, D.; Allemand, J.; Benkovic, S. J.; Croquette, V. Mechanism of Strand Displacement Synthesis by DNA Replicative Polymerases. *Nucleic Acids Res.* **2012**, *40*(13), 6174–6186.
13. Zinder, N. D.; Lederberg, J. Genetic Exchange in *Salmonella*. *J. Bacteriol.* **1952**, *64*, 679–699.
14. Ebel-Tsipis, J.; Fox, M. S.; Botstein, D. Generalized Transduction by Bacteriophage P22 in *Salmonella typhimurium* II. Mechanism of Integration of Transducing DNA. *J. Mol. Biol.* **1972**, *71*, 449–469.

15. Morse, M. L.; Lederberg, E. M.; Lederberg, J. Transductional Heterogenotes in *Escherichia coli*. *Genetics* September. **1956**, *41*(5), 758–779.

16. Oehler, S.; Eismann, E. R.; Krämer, H.; Müller-Hill, B. The Three Operators Of The Lac Operon Cooperate in Repression. *EMBO J.* **1990**, *9*(4), 973–979.

17. Busby, S., Ebright, R. H. Transcription Activation by Catabolite Activator Protein (CAP). *J. Mol. Biol.* **2001**, *293*, 199–213.

18. Siguier, P.; Edith, E.; Varani, A.; Ton-Hoang, B.; Chandler, M. Everyman's Guide to Bacterial Insertion Sequences. *Microbiol. Spec.* April **2015**, *3*(2).

19. Leslie, A. P. Transposons: The Jumping Genes. *Nat. Educat.* **2008**, *1*(1), 204.

20. Lodish, H.; Berk, A.; Zipursky, S. L., et al. *Molecular Cell Biology, 4th ed.*; W. H. Freeman: New York, 2000.

21. Lodish, et al. *Molecular Cell Biology;* W. H. Freeman: New York, 2008, pp 158–159.

22. Rajagopala, S. V.; Casjens, S.; Uetz, P. The Protein Interaction Map Of Bacteriophage Lambda. *BMC Microbiol.* **2011**, *11*, 213.

23. Grath, S. M.; van Sinderen, D., Eds.; *Bacteriophage: Genetics and Molecular Biology;* Caister Academic Press: 2007, ISBN 978-1-904455-14-1.

24. Reader, R.W.; Siminovitch, L. Lysis Defective Mutants of Bacteriophage Lambda: On the Role of the S Function in Lysis. *Virology* **1971** March, *43*(3), 623–637.

25. GATEWAY™ Cloning Technology, Manual by Life Technologies, 2003.

26. Miller, A. D.; Coffin, J. M.; Hughes, S. H.; Varmus, H. E., Eds.; *Retroviruses. Development and Applications of Retroviral Vectors;* Cold Spring Harbor: New York, 1997.

27. Cockrell, A. S.; Kafri, T. Gene Delivery by Lentivirus Vectors. *Mol. Biotechnol.* **2007**, *36*(3), 184–204.

28. Gilbert, J. R.; Wong-Staal, F. HIV-2 and SIV Vector Systems. In *Lentiviral Vector Systems for Gene Transfer;* Gary, L. B., Ed.; Georgetown, TX, 2003 (Eurekah.com).

29. Rowe, W. P.; Huebner, R. J.; Gilmore, L. K.; Parrott, R. H.; Ward, T. G. Isolation of a Cytopathogenic Agent from Human Adenoids Undergoing Spontaneous Degeneration in Tissue Culture. *Proc. Soc. Exp. Biol. Med.* **1953**, *84*(3), 570–573.

30. Harrison, S. C. Virology. Looking Inside Adenovirus. *Science* **2010**, *329*(5995), 1026–1027.

31. Springer, C. J.; Niculescu-Duvaz, I. Prodrug-activating Systems in Suicide Gene Therapy. *J. Clin. Invest.* **2000**, *105*(9), 1161–1167.

32. Alba, R.; Bosch, A.; Chillon, M. Gutless Adenovirus: Last-generation Adenovirus for Gene Therapy. *Gene Ther.* **2005**, *12*, S18–S27. DOI: 10.1038/sj.gt.3302612.

33. Baer, A.; Kehn-Hall, K. Viral Concentration Determination Through Plaque Assays: Using Traditional and Novel Overlay Systems. *J. Vis. Exp.* **2014**, *93*, e52065 (Advance online publication, http://doi.org/10.3791/52065).

34. van Wezenbeek, P. M.; Hulsebos, T. J.; Schoenmakers, J. G. Nucleotide Sequence of the Filamentous Bacteriophage M13 DNA Genome: Comparison with Phage fd. *Gene* **1980** October, *11*(1–2), 129–148.

35. Hertveldt, K.; Beliën, T.; Volckaert, G. General M13 Phage Display: M13 Phage Display in Identification and Characterization of Protein–Protein Interactions. *Methods. Mol. Biol.* **2009**, *502*, 321–39. DOI: 10.1007/978-1-60327-565-1_19.

36. Zhou, M.; Meyer, T.; Koch, S.; Koch, J.; von Briesen, H.; Benito, J. M.; Soriano, V.; Haberl, A.; Bickel, M.; Dübel, S.; Hust, M.; Dietrich, U. Identification of a New Epitope for HIV-neutralizing Antibodies in the gp41 Membrane Proximal External

Region by an Env-tailored Phage Display Library. *Eur. J. Immunol.* **2013** February, *43*(2), 499–509.

37. Li, K.; Zettlitz, K. A.; Lipianskaya, J.; Zhou, Y.; Marks, J. D.; Mallick, P.; Reiter, R. E.; Wu, A. M. A Fully Human scFv Phage Display Library for Rapid Antibody Fragment Reformatting. *Protein Eng. Des. Sel.* **2015** October, *28*(10), 307–316.

38. Jansen, R., Embden, J. D.; Gaastra, W.; Schouls, L. M. Identification of Genes that are Associated with DNA Repeats in Prokaryotes. *Mol. Microbiol.* **2002**, *43*, 1565–1575.

39. Makarova, K. S.; Grishin, N. V.; Shabalina, S. A.; Wolf, Y.; Koonin, E. V. A Putative RNA-interference-based Immune System in Prokaryotes: Computational Analysis of the Predicted Enzymatic Machinery, Functional Analogies with Eukaryotic RNAi, and Hypothetical Mechanisms of Action. *Biol. Direct.* **2006,** *1*, 7.

40. Makarova, K. S.; Haft, D. H.; Barrangou, R.; Brouns, S. J.; Charpentier, E.; Horvath, P.; Moineau, S.; Mojica, F. J.; Wolf, Y.; Yakunin, A. F., et al. Evolution and Classification of the CRISPR-Cas Systems. *Nat. Rev. Microbiol.* **2011b,** *9*, 467–477.

41. Barrangou, R.; Fremaux, C.; Deveau, H.; Richards, M.; Boyaval, P.; Moineau, S.; Romero, D. A.; Horvath, P. CRISPR Provides Acquired Resistance Against Viruses in Prokaryotes. *Science* **2007**, *315*, 1709–1712.

42. Jinek, M.; Jiang, F.; Taylor, D. W.; Sternberg, S. H.; Kaya, E.; Ma, E.; Anders, C.; Hauer, M.; Zhou, K.; Lin, S., et al. Structures of Cas9 Endonucleases Reveal RNA-mediated Conformational Activation. *Science* **2014,** *343*, 1247997.

43. Makarova, K. S.; Wolf, Y. I.; Alkhnbashi, O. S.; Costa, F.; Shah, S. A.; Saunders, S. J.; Barrangou, R.; Brouns, S. J.; Charpentier, E.; Haft, D. H.; Horvath, P.; Moineau, S.; Mojica, F. J.; Terns, R. M.; Terns, M. P.; White, M. F.; Yakunin, A. F.; Garrett, R. A.; van der Oost, J.; Backofen, R.; Koonin, E. V. An Updated Evolutionary Classification of CRISPR-Cas Systems. *Nat. Rev. Microbiol.* **2015**, *13*(11), 722–736.

44. Wiedenheft, B.; Sternberg, S. H.; Doudna, J. A. RNA-guided Genetic Silencing Systems in Bacteria and Archaea. *Nature* **2012** February, *482*(7385), 331–338.

45. Deng, L.; Garrett, R. A.; Shah, S. A.; Peng, X.; She, Q. A Novel Interference Mechanism by a Type III B CRISPR-Cmr Module in Sulfolobus. *Mol. Microbiol.* **2013**, *8*(5), 1088–1099.

46. Li, Y.; Pan, S.; Zhang, Y.; Ren, M.; Feng, M.; Peng, N.; Chen, L.; Xiang, L. Y.; She, Q. Harnessing Type I and Type III CRISPR-Cas Systems for Genome Editing. *Nucl. Acids Res.* **2015**, DOI: 10.1093/nar/gkv1044.

47. Carter, B. J. Adeno-associated Virus and Adeno-associated Virus Vectors for Gene Delivery. In DD Lassic and N Smyth Templeton, *Gene Therapy: Therapeutic Mechanisms and Strategies*; Marcel Dekker, Inc.: New York City, 2000, pp. 41–59.

48. Herskowitz, I. Life Cycle of the Budding Yeast *Saccharomyces cerevisiae. Microbiol. Rev.* **1988**, *52*(4), 536–553.

49. Engel, S. R.; Dietrich, F. S.; Fisk, D. G.; Binkley, G.; Balakrishnan, R.; Costanzo, M. C.; Cherry, J. M. The Reference Genome Sequence of *Saccharomyces cerevisiae*: Then and Now. *G3: Genes Genom. Genet.* **2014**, *4*(3), 389–398. http://doi.org/10.1534/g3.113.008995

50. Houston, P.; Simon, P. J.; Broach, J. R. The *Saccharomyces cerevisiae* Recombination Enhancer Biases Recombination During Interchromosomal Mating-type Switching but not in Interchromosomal Homologous Recombination. *Genetics* **2004,** *166*(3), 1187–1197.

51. Traven, A.; Jänicke, A.; Harrison, P.; Swaminathan, A.; Seemann, T.; Beilharz, T. H. Transcriptional Profiling of a Yeast Colony Provides New Insight into the Heterogeneity of Multicellular Fungal Communities. *PLoS One* **2012**, *7*(9), e46243. DOI: 10.1371/journal.pone.0046243.

52. Braberg, H.; Alexander, R.; Shales, M.; Xu, J.; Franks-Skiba, K. E.; Wu, Q.; Krogan, N. J. Quantitative Analysis of Triple Mutant Genetic Interactions. *Nat. Protoc.* **2014**, *9*(8), 1867–1881. http://doi.org/10.1038/nprot.2014.127

53. Ben-Aroya, S.; Pan, X.; Boeke, J. D.; Hieter, P. Making Temperature-sensitive Mutants. *Met. Enzymol.* **2010**, *470*, 181–204.

54. Miller, J.; Stagljar, I. Using the Yeast Two-hybrid System to Identify Interacting Proteins. *Met. Mol. Biol.* **2004**, *261*, 247–262.

55. Hurstel, S.; Granger-Schnarr, M.; Daune, M.; Schnarr, M. In Vitro Binding of LexA Repressor to DNA: Evidence for the Involvement of the Amino-terminal Domain. *EMBO J.* **1986**, *5*(4), 793–798.

56. Van Criekinge, W.; Beyaert, R. Yeast Two-Hybrid: State of the Art. *Biol. Proced. Online* **1999**, *2*, 1–38.

57. Kroetz, M. B.; Hochstrasser, M. Identification of SUMO-Interacting Proteins by Yeast Two-Hybrid Analysis. *Met. Mol. Biol.* **2009**, *497*, 107–120.

58. Pawlowski, A.; Riedel, K.-U.; Klipp, W.; Dreiskemper, P.; Gross, S.; Bierhoff, H.; Masepohl, B. Yeast Two-Hybrid Studies on Interaction of Proteins Involved in Regulation of Nitrogen Fixation in the Phototrophic Bacterium *Rhodobacter capsulatus*. *J. Bacteriol.* **2003**, *185*(17), 5240–5247.

59. Lin, Y.; Li, Y.; Zhu, Y.; Zhang, J.; Li, Y.; Liu, X.; Si, S. Identification of Antituberculosis Agents That Target Ribosomal Protein Interactions Using a Yeast Two-hybrid System. *Proc. Natl. Acad. Sci. USA* **2012**, *109*(43), 17412–17417.

60. Botstein, D.; Chervitz, S. A.; Cherry, J. M. Yeast as a Model Organism. *Science* (New York, N.Y.) **1997**, *277*(5330), 1259–1260.

61. Shao, Z.; Zhao, H. DNA Assembler, an In Vivo Genetic Method for Rapid Construction of Biochemical Pathways. *Nucleic Acids Res.* **2009**, *37*(2), e16. http://doi.org/10.1093/nar/gkn991

62. Mattanovich, D.; Branduardi, P.; Dato, L.; Gasser, B.; Sauer, M.; Porro D. Recombinant Protein Production in Yeasts. *Met. Mol. Biol.* **2012**, *824*, 329–358.

63. Da Silva, N. A.; Srikrishnan, S. Introduction and Expression of Genes for Metabolic Engineering Applications in *Saccharomyces cerevisiae*. *FEMS Yeast Res.* **2012**, *12*, 197–214.

64. Rajbhandary, U. L. More Surprises in Translation: Initiation Without the Initiator tRNA. *Proc. Natl. Acad. Sci. USA* **2000**, *97*(4), 1325–1327.

65. Ruohonen, L.; Aalto, M. K.; Keränen, S. Modifications to the ADH1 Promoter of *Saccharomyces cerevisiae* for Efficient Production of Heterologous Proteins. *J. Biotechnol.* **1995**, *39*(3), 193–203.

66. Partow, S.; Siewers, V.; Bjørn, S.; Nielsen, J.; Maury, J. Characterization of Different Promoters for Designing a New Expression Vector in *Saccharomyces cerevisiae*. *Yeast* **2010**, *27*(11), 955–964. DOI: 10.1002/yea.1806.

67. Nevoigt, E.; Kohnke, J.; Fischer, C.R.; Alper, H.; Stahl, U.; Stephanopoulos, G. Engineering of Promoter Replacement Cassettes for Fine-tuning of Gene Expression in *Saccharomyces cerevisiae*. *Appl. Environ. Microbiol.* **2006**, *72*, 85266–85273.

68. Widlund, P. O.; Davis, T. N. A High-efficiency Method to Replace Essential Genes with Mutant Alleles in Yeast. *Yeast.* **2005,** *22*(10), 769–774.

69. Sikorski, R. S.; Hieter, P. A System of Shuttle Vectors and Yeast Host Strains Designed for Efficient Manipulation of DNA in *Saccharomyces cerevisiae. Genetics* **1989,** *122*(1), 19–27.

70. Voth, W. P.; Richards, J. D.; Shaw, J. M.; Stillman, D. J. Yeast Vectors for Integration at the *HO* Locus. *Nucleic Acids Res.* **2001,** *29*(12), e59.

71. MacAlpine, D. M.; Perlman, P. S.; Butow, R. A. The Numbers of Individual Mito-chondrial DNA Molecules and Mitochondrial DNA Nucleoids in Yeast are Co-regu-lated by the General Amino Acid Control Pathway. *EMBO J.* **2000,** *19*(4), 767–775.

72. Goldring, E. S.; Grossman, L. I.; Marmur, J. Petite Mutation in Yeast. *J. Bacteriol.* **1971,** *107*(1), 377–381.

CHAPTER 4

PLANT MOLECULAR BIOLOGY

CONTENTS

ABSTRACT

As humans evolved from nomadic to agriculture-based societies, they utilized "special" techniques to modify plants and animals. Some of the food crops such as rice were converted from being perennial to annual plants. Moreover, the traits that were of most value to humans and domesticated animals have been selected for, even without the knowledge of "genes" and recombinant DNA technology. Although the plant genomes are very large, the genomes can be compared with one another by mapping the locations of certain genes or gene traits in various plants. Whole genome sequencing has helped in the discovery of genomic variations and genes associated with adaptation to climatic changes. Significant genomic advances have been made for abiotic stress tolerance in plants with the help of special techniques. In this chapter, we will also briefly discuss other molecular components of signaling pathways, the crosstalk among various abiotic stress responses, and use in improving abiotic stress tolerance in different crops.

4.1　EARLY APPROACHES

Humans have actually been genetic engineers for thousands of years. As humans evolved from nomadic to agriculture-based societies, they utilized the "genetic engineering" techniques to modify plants and animals—bringing about changes in the gene pool within crop species. For example, in maize and wheat, the trait of seed dispersal was selected against, thus making these plants completely dependent on humans for seed dispersal. In addition, some of the food crops such as rice were converted from being perennial to annual plants. Moreover, increased size of plant parts such as fruits, storage organs, roots, etc. that were of most value to humans and domesticated animals have been selected for. These changes were carried out by selecting and propagating individuals with the desired traits, even without the knowledge of "genes" and recombinant DNA technology.

In the last century, a growing understanding of genetics helped in the rate of crop improvement. However, increased inbreeding led to decreased yields because the deleterious genes too became homozygous. One of the most outstanding agricultural achievements was the development of hybrid corn with increased "hybrid vigor."[1] This was achieved by crossing two different inbred lines giving rise to hybrid offspring that was highly productive, as in the case of hybrid wheat (Fig. 4.1). Hybrid rice developed by the International Rice Research Institute in the Philippines has increased yield 20%.[2] Another approach to optimize food quality is by targeting specific genes, a field that holds a lot of promise since only a small percentage of the genes and their function have been identified, but this century has witnessed a lot of advancements in technologically powerful new ways to understand genomes.

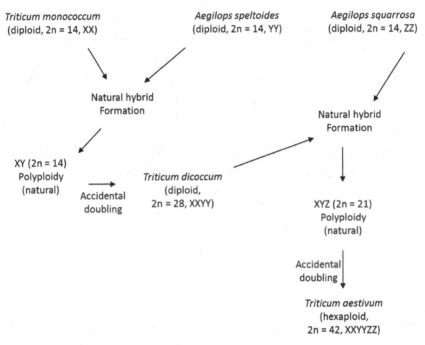

FIGURE 4.1   Evolutionary history of wheat.

4.1.1   *ORGANIZATION OF PLANT GENOMES*

The genomes of plants are more complex than that of other eukaryotes; their analysis reveals many evolutionary changes in the DNA sequences over time. Plants show widely different chromosome numbers and varied ploidy levels (Fig. 4.2). Overall, the size of plant genomes (both number of chromosomes and total nucleotide base-pairs) exhibits one of the greatest variation of any kingdom. For example, the genome size of members of genus *Triticum* contain nearly over 120 times as much DNA as the small weed *Arabidopsis thaliana* (Table 4.1). The DNA of plants, similar to animals, can also contain regions of sequence repeats, insertion elements, or sequence inversions, which further modify their genetic content. Increasingly, researchers are turning to studying the organization of plant DNA sequences to obtain important information about the evolutionary history of a plant species.

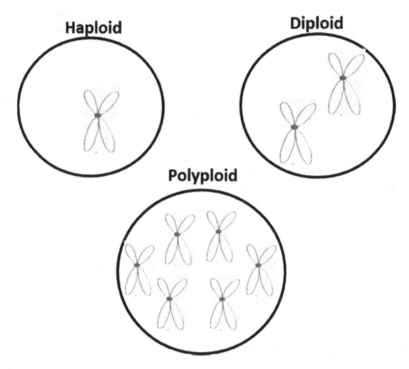

FIGURE 4.2 Different levels of ploidy in plants. Cells are described according to the number of chromosomal sets present. Shown here are monoploid (1 set), diploid (2 sets), and polyploid (many sets).

4.1.2   LOW-, MEDIUM-, AND HIGH-COPY-NUMBER DNA

In most seed plants, a very small percentage of the genome actually encode genes involved in the production of protein and are often referred to as "low-copy-number DNA." It has been seen that most of these

sequence alterations occur in noncoding regions. An important component of the cellular machinery, ribosomal RNA (rRNA), that translates
transcribed messenger RNA (mRNA) into protein are encoded by the
DNA sequences that are known as "medium-copy-number DNA." rRNA
genes may be present in several hundred to several thousand copies in
plant genomes, in contrast to animal cells, where only 100–200 rRNA
genes are normally present. The evolutionary patterns of plant species
can be analyzed by the degree of variations in plant genomes with
respect to the number and mutational analysis of their rRNA genes.
Plant genomes may also contain highly repetitive sequences, or "high-
copy-number DNA." The function of these high-copy-number DNA in
plant genomes is still waiting to be discovered. Roughly half the maize
genome is composed of such DNA.

4.1.3 SEQUENCE REPLICATION AND INVERSION

There is a lack of correlation between complexity and size of eukaryotic genomes, largely due to the presence of noncoding highly repetitive DNA. This phenomenon is commonly observed in higher plants. It
is also observed that the protein-coding sequences in the genomes are
generally similar in different plant species, and that the repetitive DNA
mainly account for the variation in genome size (Fig. 4.3). These repetitive sequences have accumulated in the genomes in the evolutionary
process.

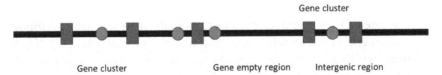

FIGURE 4.3   Genes are present in gene-rich regions isolated with long regions of repetitive
DNA.

The high-copy repetitive DNA sequences may be organized in different
possible combinations within a plant genome[3] (Fig. 4.4). Several copies
of a single repetitive DNA sequence may be present in "simple tandem
array" together in the same orientation. Alternatively, these sequences can

be spread within single-copy DNA in a same orientation as "repeat/single-copy interspersion," or in the opposite orientation as "inverted repeats." Other possible arrangements of groups of repetitive DNA sequences, in plant genomes are the "compound tandem array" or a "repeat/repeat interspersion."

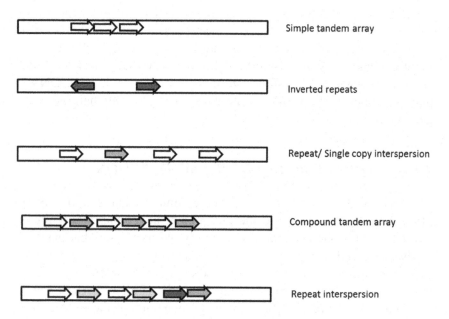

FIGURE 4.4 Organization of repeated sequences in the genome. Direction of arrow shows sequence orientation while same shade indicates similar sequence.

Clustered DNA repeats are transcriptionally inert and can be found in centromeric and telomeric heterochromatin. For example, *CENH3, a* centromeric DNA is the most abundant tandem repeat, and is found in both plants and animals. Other characteristics of repetitive sequences are:

(a) Consistent presence of motifs such as AA/TT dinucleotides, pentanucleotide CAAAA, etc. in different families of repetitive sequences.

(b) A characteristic feature of various plant satellite families is the presence of short, direct and inverted repeats, and short palindromes.

These palindromes may act as preferred sites for rearrangements by acting as potential substrates for homologous recombination.

(c) Methylation is another characteristic feature of repetitive sequences.

A few repetitive species in different plant species are[3]: onion (e.g., ACSAT 1/2/3); *Arabidopsis* (e.g., 180 bp repeat/*Hind*III repeat/AtCen/ pAL1/pAS1/pAtMR/pAtHR/pAa214/AaKB27 family); tomato (e.g., GR1 and pLEG15); tobacco (e.g., HRS60 and TAS49); rice (e.g., C154, C193, OsG5, TrsC, and CentO-C, etc.); maize (e.g., Cent4, MR68, MR77, and CentC).

The high-copy repetitive DNA sequences may be organized in different possible combinations within a plant genome. The presence of repetitive DNA can vastly increase the plant genome size, making it difficult to find and characterize individual single-copy genes. The presence of highly repetitive DNA sequences in plant genomes can be explained by a variety of mechanisms. Repetitive sequences can be generated by DNA sequence amplification in which multiple rounds of DNA replication occur for specific chromosomal regions. Unequal crossing over of the chromosomes during meiosis or mitosis (translocation) or the action of transposable elements (see next section) can also generate repetitive sequences.

Next-generation sequencing (NGS) technologies have helped in gaining more information about repetitive sequences. By applying NGS technologies to very complex populations of plant repetitive elements, it has been possible to characterize genomes and establish phylogenies in species. Various strategies such as single nucleotide polymorphism (SNP) detection and other approaches are being developed to analyze repeats and to assemble NGS data to help in understanding their role in gene function and evolution.[4] In addition, most abundant tandem repeats from diverse plant and animal were identified through whole genome shotgun sequencing.[5]

Several web-based tools such as REViewer, RepEx, and RepeatExplorer have been developed for analyzing repetitive sequences.

A major limitation in studying repetitive sequences is that their cloning and sequencing is technically challenging, hence, approaches such as mapping and sequence analysis are also applied. These sequences also pose challenges in sequencing and assembling of genomes. Thus, genome-wide analysis, whole genome resequencing, transposon-based sequencing, and fine mapping of repetitive sequences can elucidate the structure, evolution, and functional potential of these yet not fully studied components of a complex genome.

4.1.4 TRANSPOSABLE ELEMENTS

As discussed in Chapter 3, these are special sequences of DNA with the ability to move from place to place in the genome. These elements are also called "jumping genes" because they can excise from one site at and reinsert in another site. Transposable elements often insert into coding regions or regulatory regions of a gene, thus affecting expression of that gene, resulting in a mutation that may or may not be detectable (Figs. 3.8 and 3.9; Chapter 3). In 1950, Barbara McClintock studied transposable elements in corn, which led her to win the Nobel Prize in 1983 for her work.[6] Transposable elements can also be involved in generating repetitive DNA sequences because they can move through the genome and their capacity to replicate independently. This is believed to be the case with the extensive retroviral-like insertions in maize (Fig. 4.5). In addition, each instance of repetitive sequence insertion might involve a mutation in the transposable element itself which removes its capacity to transpose and be retained in that site in the genome.

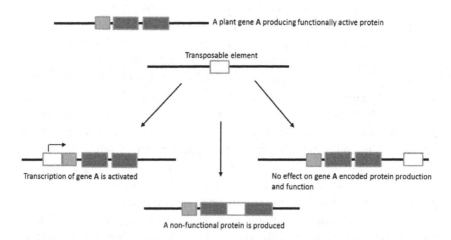

FIGURE 4.5 Different kinds of transposition in plants. Effects of movement of a transposable element on the target gene expression. The transposable element is shown in light grey, and the target gene (A) is composed of multiple exons. Protein coding regions of exons are dark grey and untranslated regions are light grey. The perpendicular arrow (⌐►) indicates the start site for transcription.

4.2   PLANT GENOME PROJECTS

The plant genome projects address the great potential of plants of economic importance on a genome-wide scale. There has been a tremendous increase in the availability of functional genomics tools and sequence resources for use in the study of key crop plants and their models. Expert research teams from all over the world are focusing on: addressing fundamental questions in plant sciences on a genome-wide scale and not limited to genes only; and developing resources such as databases and tools for plant genome research and analysis.

The potential of having complete genomic sequences of plants is tremendous and about to be realized now that a few plant genomes have been completely sequenced (Table 4.1). The completely sequenced genomes will have far-reaching uses in agricultural breeding and evolutionary analysis. In plant genomes, the gene order seems to be more conserved than the nucleotide sequences of homologous genes. In grasses used by humans for grain production, differences in genome size can largely be attributed to different quantities of inserted LTR transposons.[7,8] Sequencing the rice genome provides a model for a small monocot genome. Rice was selected, in part, because its genome is 6, 10, and 40 times smaller than maize, barley, and wheat (Table 4.1). These grains represent a major food source for humans. The understanding of rice genome has made it much easier to study the grains with larger genomes. Even though these plants diverged more than 50 million years ago, the chromosomes of rice, corn, barley, wheat, and other grass crops show extensive conserved arrangements of segments[8] (synteny) (Fig. 4.6). DNA sequence analysis of cereal grains will be important for identifying genes associated with growth capacity, yield, nutritional quality, and disease resistance.

TABLE 4.1   Comparison of Different Plant Genome Sizes

Plant	Genome size (Mbp)	Number of genes[ref]
Oriza sativa	374.55	~40,464[18]
Triticum aestivum	15,966	>124, 201[19]
Lycopersicum esculentum	907	~34,727[20]
Zea mays	2500	~40,000[21]
Arabidopsis thaliana	135	27,655[22]
Glycine max	950	46,430[17]

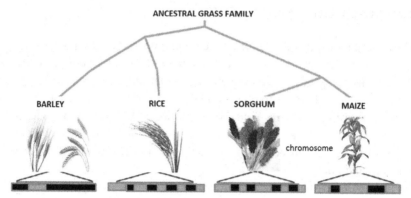

FIGURE 4.6   Synteny can be observed in grass family. Significant similarity in the gene content of different grass species is observed when the grass genomes are mapped by using common sets of low-copy-number DNA markers. The difference in genome size is attributable mainly to differences in number of repetitive DNA. Grass species show great variations in genome size and chromosome number.

4.2.1   PLANT FUNCTIONAL GENOMICS AND PROTEOMICS

Arabidopsis and rice genome sequencing represent major technological accomplishments. Bioinformatic studies use high-end technology to analyze the growing gene databases, look for phylogenetic relationships among genomes, and hypothesize functions of genes based on sequence analyses. International community of researchers has come together to study the function of many plant genomes. One of the first steps is to determine the spatial and temporal regulation of these genes. Each step beyond that will require additional enabling technology. Research will move from genomics to proteomics (the study of all proteins in an organism). Proteins are much more difficult to study because of posttranslational modification and formation of complexes of proteins. The information obtained will be essential in understanding physiology, cell biology, development, and evolution. For example, how are similar genes used in different plants to create biochemically and morphologically distinct organisms? So, in many ways, we continue to ask the same questions that even Mendel asked, but at a much different level of organization.

The observation that the genome components of rice, wheat, sugar cane, and corn are highly conserved implies that the order of the segments in the ancestral grass genome has been rearranged by recombination leading to the evolution of the grasses.[9]

4.2.2 PLANT COMPARATIVE GENOME MAPPING

Traditionally, plant molecular phylogenetics involves amplifying, sequencing, and analyzing genes from many species. At the same time, the advent of new techniques to study DNA sequences, such as NGS, gene mapping, and chromosome synteny has helped in studying plant genomes. NGS technologies allow mass sequencing of genomes and transcriptomes, and produce a vast line-up of information. The analysis of NGS data helps in discovering new genes, regulatory sequences and their positions, and makes available large collections of molecular markers. With an increase in understanding of plant genomes, better manipulation of genetic traits such as crop yield, disease resistance, growth abilities, nutritive qualities, and stress tolerance can be practiced. Each of these traits is encoded by sets or multiple genes. Some mechanisms and processes conserved across the plant kingdom can be studied on any model species, while others have evolutionarily diverged and can be studied only on closely related model species. *Arabidopsis* and rice species have been adopted as models for dicotyledons and monocotyledons and more recently brachypodium.[10] These model plants are diploids, have rapid life cycles, well-developed genetics, fewer and smaller chromosomes, and are easily transformed. Moreover, these models have their technical resource databases curated by international centers. Moreover, genomes of model species share significant genetic synteny with important crop plants and facilitate gene discoveries and subsequently their phenotypic association.

4.2.3 TOOLS TO MAP GENOMES AND DETECT POLYMORPHISMS

Molecular marker techniques are helpful to elucidate stress related traits by quantitative trait locus (QTL) mapping in order to locate the individual loci through marker-assisted selection. In the classical approach, a linkage map is made by calculating the frequency of recombination. The map positions are inferred from estimates of recombination frequencies between genes. The frequency of recombination is used to calculate distance[11], and subsequently, the linkage map. However, this approach can be applied to genes with alleles that can be phenotypically identified.

4.2.3.1   RESTRICTION FRAGMENT LENGTH POLYMORPHISMS

Restriction fragment length polymorphisms (RFLP) involves analysis of the pattern of DNA fragments, produced when DNA is treated with restriction enzymes. RFLP takes the advantage of polymorphisms in individual's genotype that give rise to variations in phenotype. If the location of a particular gene corresponding to a trait is being mapped in a certain chromosome, the DNA of members of that species with the trait is analyzed, and similar patterns of inheritance in RFLP alleles are searched. Once a specific gene is localized, conducting RFLP analysis on other members of the species could reveal a carrier of the mutant genes. Thus, RFLPs are fragments of DNA that may contain a part of one or more genes. In addition, the RFLP analysis technique is tedious and slow. Besides requiring a large amount of DNA, and a suitable probe library, the whole procedure process could take up to a full month to complete. Currently, the very dense RFLP map is in rice where 2000 DNA sequences have been mapped onto 12 chromosomes (Fig. 4.7).[12]

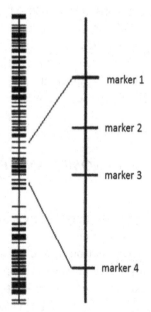

FIGURE 4.7   Representation of RFLP map of a chromosome of rice. Horizontal lines depict specific markers genes and are placed according to their respective distances. The distance between the genetic markers is mentioned in centimorgans.

4.2.3.2   AMPLIFIED FRAGMENT LENGTH POLYMORPHISM

Amplified fragment length polymorphism (AFLP) is another technique that utilizes genome sequence variability. The DNA fragments that have been cut with restriction enzymes, (usually *EcoRI* and *MseI*), are hybridized with DNA primers and subsequently amplified using the polymerase chain reaction (PCR) to generate AFLP maps.[13] Many fragment subsets can be amplified by changing the nucleotide extensions on the adapter sequences. Thus, hundreds of markers can be generated reliably. The resulting PCR products, representing pieces of DNA cut by a restriction enzyme, are separated by gel electrophoresis. The band sizes on an AFLP gel tend to show more polymorphisms than those found with RFLP mapping because the entire genome is visible on the gel and a high resolution is obtained because of stringent PCR conditions. Both RFLPs and AFLPs (among many other tools for genome analysis) can provide markers of traits which are inherited from parents to progeny through crosses.

4.2.3.3   SIMPLE SEQUENCE REPEATS OR MICROSATELLITES

These are tandem repeats of one to six nucleotides, and are considered important because they are reproducibile, hypervariabile, relatively abundant, multiallelic, cover genome extensively, and are amenable to high throughput genotyping through automation. Microsatellites (simple sequence repeats, SSRs, as shown in Fig. 4.8) occur frequently in most eukaryote genomes[13], and can be either developed from genomic DNA libraries or from enriched libraries for specific microsatellites. These can also be found by searching GenBank, EMBL and other public databases. EST databases provide an valuable source of potential genes, as these can generate markers directly associated with a trait of interest and may be transferred and checked in a related genera. In case the nucleotide sequence of the flanking regions of the microsatellites are known, primers can be designed and the polymorphisms can be detected by southern hybridization or by PCR. This technique can amplify large number of DNA fragments per reaction representing multiple loci across the genome.

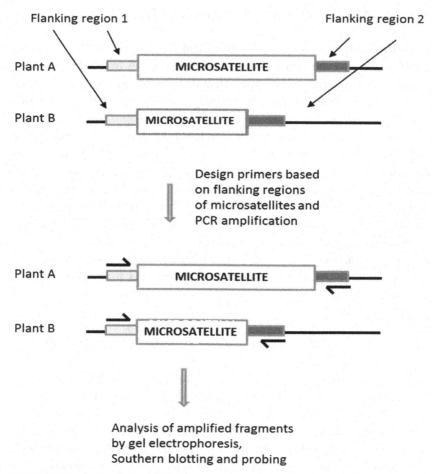

FIGURE 4.8   A schematic of SSR assay.

4.2.3.4 MICROARRAYS

DNA microarray helps to relate sequences with the study of gene function. Also, called biochips or genes-on-chips, these assays enable the study of the presence of a particular stage of a gene. To prepare a particular DNA microarray, fragments of DNA of the organism grown under certain conditions are mechanically deposited on a microscope slide at indexed locations. Nearly 10,000 spots can be displayed over an area of only 3.24 cm^2 (Fig. 4.9). Microarrays primarily help to determine the genes that are expressed developmentally in specific tissues or in response to certain environmental factors. The microarrays are then probed with RNA isolated from these tissues and only those sequences that are expressed in the tissues will be present to hybridize with the specific spot on the microarray.

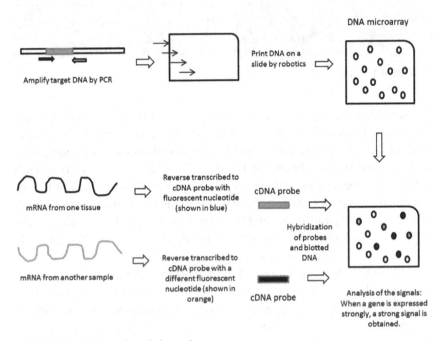

FIGURE 4.9   Schematics of plant microarrays.

4.2.4   A. THALIANA AS A MODEL SYSTEM FOR PLANT GENOME ANALYSIS

A. thaliana (Fig. 4.10) is a member of the mustard family (Cruciferae or Brassicaceae). Although it is not significant agriculturally, *Arabidopsis* offers important information for research in plant genetics and molecular biology.[14] First, nearly everything, in terms of size, about *Arabidopsis* is small, including its entire life cycle which is completed in 6–8 weeks. Bolting starts at about 3 weeks after planting, and the resulting inflorescence forms a linear progression of flowers and siliques for several weeks before the onset of senescence.

FIGURE 4.10   Photographs of *Arabidopsis*.

Second, mature plants often produce several hundred siliques with very prolific seed production. Third, flowers are very small (2 mm long) and can self-pollinate as the bud opens, and be cross-pollinated by applying pollen to the stigma surface. Fourth, small unicellular hairs known as trichomes cover the leaves and are convenient models for studying cellular differentiation and morphogenesis. The roots are simple in structure, easy to study in culture. The plant is amenable to be transformed by *Agrobacterium tumefaciens*. Finally, plants can be grown in petri plates or maintained in limited space such as pots in a greenhouse.

The tremendous development in *Arabidopsis* research over the last three decades have further increased its utility for molecular genetics.[14] The genome has been sequenced and annotated and extensive genetic and physical maps of all chromosomes are available (*Arabidopsis* Genome Initiative AGI, 2000). Since the plant has a diploid genome, recessive mutations can be easily analyzed.[15]

Over 330,000 insertions (resulting in the loss of function of the gene product) in virtually all *Arabidopsis* genes have been created and identified at precisely sequenced locations. Repertoire of gene families in *Arabidopsis* (11,000–15,000) is similar to other sequenced multicellular eukaryotes. However, gene number in *Arabidopsis* is surprisingly high—nearly 30,700 genes. Some of these extra genes are due to genome duplications. Nearly 8000 (25%) of *Arabidopsis* genes have homologs in the rice genome, but not in drosophila, *C. elegans,* or yeast. Over 81% of ORFs fall within the bounds of a block, whereas only 28% of genes are present in duplicate due to extensive deletions of genes.

4.2.5   GENOME SEQUENCING OF RICE AND OTHER GRAINS

Rice (*Oryza sativa*) is the principal food of half of the world's population. It has been a decade since complete genome sequencing of rice conducted by the International Rice Genome Sequencing Project (IRGSP) has been achieved. Rice was the first completely sequenced crop genome, paving the way for the sequencing of more complex crop genomes.

The genome sequence made an immediate impact on rice genetics and breeding research, as evidence by the use of DNA marker and citations. The impact on other crop genomes, particularly for those within the grass family was evident too.

Rice is a model cereal plant (Fig. 4.11) for research because of the small size of its genome (~375 Mb), its relatively short generation time, its relative genetic simplicity (it is diploid, or has two copies of each chromosome), its ease to transform genetically, and it belongs to the grass family which has the greatest biodiversity of cereal crops.[16] The rice plant has a high degree of collinearity with the genomes of wheat, barley, and maize.

FIGURE 4.11   Photograph of rice plants.

The sequenced segment of rice genome represents 99% of euchromatin and 95% of rice genome. The rice genome has nearly 37,344 coding genes. One gene can be found every 9.9 kb, a lower density than that observed in *Arabidopsis*. Nearly 2859 genes are unique to rice and other cereals. Repetitive DNA is estimated to constitute at least 50% of the rice genome.

The sequencing of rice genome has many far-fetched results. First, development of gene-specific markers for marker-assisted breeding of new and improved rice varieties is now more feasible. Second, it is easier to understand how a plant responds to the environment and which genes control various functions of the plant. Third, with the sequence analyses, the nutritional value of rice can be improved and crop yield can be enhanced by improving seed quality. Finally, the sequence information is useful in identifying plant-specific genes that can be potential herbicide targets.

The soybean (*Glycine max*) genome was published in 2010.[17] It also took a long time because it is relatively large at around 1 Gbp with numerous transposons, and lot of duplicated genes. One of the most important features of soybeans is their production of lipids, with soybean oil being one of the major products. They tried to annotate all the genes possibly involved in lipid metabolism, and came up with 1157.

4.2.6 CHLOROPLAST GENOME AND ITS EVOLUTION

The chloroplast is a plant organelle that plays important role in photosynthesis, and can replicate independently in the plant cell. Plant chloroplasts have their own specific DNA, which is independent of that present in the nucleus. The chloroplast DNA is maternally inherited and encodes proteins unique to the chloroplast. Many of the proteins encoded by chloroplast DNA are involved in photosynthesis. These characteristics give rise to the hypothesis that chloroplasts could have originated from a photosynthetic prokaryote that became part of a plant cell by endosymbiosis. Many prokaryote-like features have been observed in the chloroplast DNA, similar to their double-stranded circular loops, like that of prokaryotic chromosomal DNA. In addition, chloroplast DNA also contains genes for ribosomes that are very similar to those present in prokaryotes. The order of assembly and number of genes in chloroplast DNA of all land plants is nearly the same (~100), (Fig. 4.12).[23] The chloroplast DNA, compared to the plant nuclear DNA has evolved at a more conservative pace, and therefore shows a more interpretable evolutionary pattern when scientists study DNA sequence similarities. Moreover, chloroplast DNA is also not subject to recombination-induced mutations and modification caused by transposable elements.[24] In the evolutionary history, some genetic exchange between the nuclear and chloroplast genomes appears to have taken place. For example, the key enzyme (RUBISCO) in the Calvin cycle of photosynthesis consists of a large and small subunit. The small subunit is encoded by the nuclear genome. The protein it encodes has a targeting sequence that allows it to enter the chloroplast and combine with large subunits.

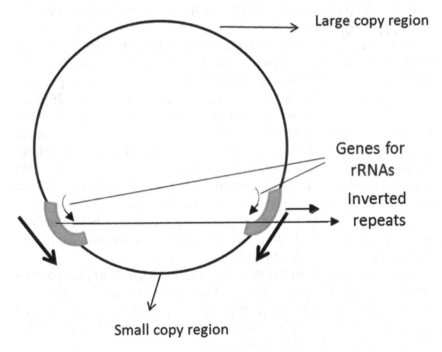

FIGURE 4.12   A chloroplast genome.

Another characteristic feature of the chloroplast DNA is the presence of two nearly identical inverted repeats,[25] whose length may vary from

4000 to 25,000 base pairs. The inverted repeat regions usually contain tRNA and ribosomal RNA genes. While a given pair of inverted repeats are rarely completely identical, they are always very similar to each other, and are highly conserved among land plants, and accumulate few mutations.[26] The genomes of cyanobacteria and that of glaucophyta and rhodophyceae contain similar inverted repeats, suggesting that they predate the chloroplast. However, some chloroplast DNAs have lost the inverted repeats, similar to those of peas and a few red algae;[27] have one of their inverted repeats flipped, making them direct repeats, similar to that of red alga *porphyra*.[28] It has been observed that the chloroplast DNAs which have lost some of the inverted repeat segments tend to get rearranged more, and possibly help stabilize the rest of the chloroplast genome.[27] Other DNA sequence inversions or deletions occur rarely, for instance, a large inversion in chloroplast DNA is found in the Asteraceae, or sunflower family, and not in other plant families.

There is increasing use of plant molecular data such as chloroplast DNA sequences. Sequence information on chloroplasts is available on the ChloroMitoSSRDB database which currently provides access to 2161 organellar genomes (1982 mitochondrial and 179 chloroplast genomes).[29] Proteins found in at least one plastid genome have nucleus-encoded counterparts in other species.[30] Following the initial publications, predicting the size and evolutionary origin of the chloroplast proteome encoded in *A. thaliana* predicted nearly 1900 and 2500 nucleus-encoded chloroplast proteins, of which a minimum of 35% derived from the cyanobacterial ancestor. When considered together, a clearer understanding of the evolutionary processes can be obtained from the morphological and molecular information and provide factors that govern biological diversity.

It is also observed that on comparisons of predicted chloroplast proteins sets between *Arabidopsis* and rice defined a subset of around 900 tentative chloroplast proteins, predominantly derived from the cyanobacterial endosymbiont with function mostly related to transcription, metabolism, and energy that is shared by both species.[31]

The chloroplast DNA replicates using a double-displacement loop (D-loop)[32], or through replication structures similar to bacteriophage T4 in cases of linear chloroplast DNA. A theta intermediary form is made as the D-loop moves through the circular DNA, and uses a rolling circle mechanism to complete the replication process. Multiple replication forks open up, allowing replication machinery to transcribe the DNA. As replication continues, the forks grow and eventually converge. The daughter chloroplast DNA structures separate, creating daughter chloroplast DNA.

4.3 PLANT TRANSFORMATION

There are several methods to insert or transform foreign genes into plants. For this, the DNA fragment coding for the protein of interest and an associated promoter whose expression is targeted to a particular stage of developmental or tissue and is integrated into the genome of the plant. Thus, when the plant is propagated, each plant will transmit this property to its progeny and large numbers of plants containing the transferred gene are readily generated.

Genes have also been delivered into the genome of plastids (chloroplasts and mitochondria) in plant cells. While the chloroplasts in tobacco and potato plants have been successfully transformed, research is being done to extrapolate the method to other crops. A major advantage of chloroplast transformation is that the genes in chloroplast genomes are not transmitted through pollen; recombinant genes are easier to contain, thereby avoiding unwanted escape into the environment.

A second method involves the use of a recombinant plant virus to engineer plant protein expression through transduction to deliver genes into plant cells. The DNA coding for the desired protein is engineered into the genome of a plant virus that will infect a host plant. For this, the host plants is grown till a proper stage and then inoculated with the engineered virus. As the virus replicates and spreads within the plant, within a short time, many copies of the desired DNA are made and as a result, high level of protein is produced. A limitation with this system is that the green plant matter must be processed immediately after harvest and cannot be stored.[33]

Nowadays, particle bombardment and *A. tumefaciens*-mediated transformation procedure are preferred because they can successfully transform various plant tissues such as roots and leaves, which are more stable and easier to handle.

The transformed plant cells in which genes coding is desired have been stably introduced to give the plant a new trait. These genes are flanked by promoter and terminator regions (often a 35S cassette; *ubi* in the mono-cots) that are recognized by the plant transcription machinery. Thus, the genes can express the desired protein. Transformed plant cells can grow on selective media containing an antibiotic (kanamycin, hygromycin, etc.) or a herbicide (phosphinothricin) because the transformation vectors contain genes that encode resistance properties.

The naturally occurring *A. tumefaciensonc* genes that are naturally present between the 25 bp repeats of the T-DNA can be removed by dele-tion. As a result, any gene introduced between these repeats can be trans-ferred into plant cells through *A. tumefaciens,* which can be applied not only in dicotyledonous plants but also in monocots. The integration of (T)-DNA occurs at random sites in the plant genome by using either a site-specific nuclease (e.g., a zinc-finger nuclease) in homology-directed integration or site-specific recombination system (e.g., Cre-*lox*).[34] The transformed cells can have either a single copy or multicopies of the trans-gene, the latter might exhibit RNA interference, or posttranscriptional gene silencing (PTGS, see Chapter 2 for details).

However, while producing edible plant products, selection markers are not desired in mature plants. The European Union suggests avoiding the use of selectable markers in genetically engineered (GE) crops. This would cater not only to the safety concerns of GE crops but also support multiple transformation cycles for transgene pyramiding.

4.3.1 PLANT TRANSFORMATION USING THE PARTICLE GUN

This process involves using a "gun" to blast plant cells with microscopic gold particles coated with the foreign DNA at high velocity, which then is integrated into the plant genome. It can be achieved by a burst of high-pressure helium gas or an electrical discharge helps accelerate the particles to a sufficient velocity to pass through the plant cell wall. These cells are identified with the help of a selectable marker also present on the foreign DNA, to allow only those cells receiving the foreign DNA to survive on a particular growth medium (Fig. 4.13). The selectable markers include genes for resistance to herbicide or antibiotic. Plant cells which survive growth in the selection medium are then tested for the presence of the foreign gene(s) of interest.

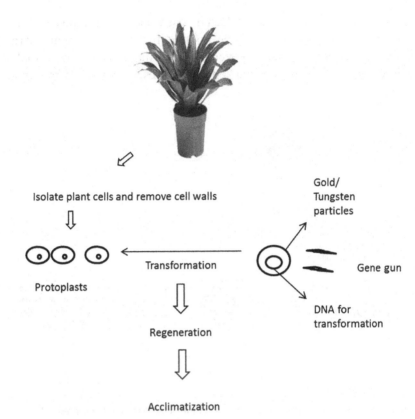

Isolate plant cells and remove cell walls

Protoplasts

Transformation

Gold/ Tungsten particles

Gene gun

DNA for transformation

Regeneration

Acclimatization

tested for the presence of the foreign gene(s) of interest.

FIGURE 4.13   Flowchart showing plant transformation using particle gun.

4.3.2   PLANT TRANSFORMATION USING ELECTROPORATION

The foreign DNA can also be sent through electrical shock into cells that lack a cell wall, such as the plant protoplasts described earlier. A pulse of high-voltage electricity briefly opens up small pores in the protoplasts' plasma membranes, allowing the foreign DNA in a solution containing plant protoplasts and foreign DNA to enter the cell. Following electroporation, the protoplasts are transferred to a growth medium for cell wall regeneration, cell division, and, eventually, the regeneration of whole plants (Fig. 4.14). The DNA incorporates into one of the plant's chromosomes. A selectable marker present in the foreign DNA and protoplasts containing foreign DNA are selected based upon their ability to survive and proliferate in a growth medium containing the selected treatment (antibiotic or herbicide). Once regenerated from the transformed protoplasts, whole plants can then be evaluated for the presence of the desired trait.

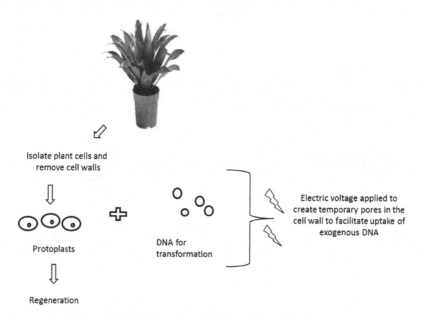

Isolate plant cells and
remove cell walls

Protoplasts

DNA for
transformation

Electric voltage applied to
create temporary pores in the
cell wall to facilitate uptake of
exogenous DNA

Regeneration

treatment (antibiotic or herbicide). Once regenerated from the transformed protoplasts, whole plants can then be evaluated for the presence of the desired trait.

FIGURE 4.14   Plant transformation using electroporation.

It becomes vital to know the localization of the protein of interest within specific plant cells or tissues for a better understanding of the factors controlling stability and accumulation of the heterologous proteins, as well as the effect of environmental conditions on this localization during development.

4.3.3   MARKER ELIMINATION STRATEGIES

Co-transformation of genes of interest along with selectable marker genes and the segregation of the separate genes through sexual crosses is one of the simplest marker elimination strategies.[35] A few co-transformation strategies can be accomplished by co-inoculating plant cells with two *Agrobacteruim* strains, each containing a simple binary vector, dual binary vector systems, and modified two-border *Agrobacterium* transformation vectors.

Selection through Ipt involves the isopentenyl transferase (*ipt*) gene that results in overproduction of cytokinine. This cytokine overproduction leads to abnormal shoot morphology in the transgenic shoots which can also be used as a selectable marker. The appearance of phenotypically normal plants emerging from abnormal tissues indicates excision of the *ipt* gene, resulting in marker-free plants.

Using "shooter" mutant Agrobacterium strains are also efficient transformation systems.[36] These mutant strains possess defective auxin-synthesis genes, but the presence of intact *ipt* gene on the T-DNA of their Ti plasmid results in transgenic cell proliferation and formation of adventitious shoots. Regeneration on growth regulator-free media only occurs after successful infection of the plant tissues by *Agrobacterium* "shooter" strain.

Use of the nuclear-encoded, plastid-targeted phage site-specific recombinases is another strategy to generate marker-free transgenic plant. Under the control of inducible promoters, the marker genes are excised. The Cre/*lox*, FLP/*FRT*, or R/*Rs* systems have been reported to be successful in different plant species in which Cre, FLP, and R are the recombinases, and *lox*, *FRT*, and *Rs* are the recombination sites, respectively.[37]

4.4 PLANT TISSUE CULTURE: AN IMPORTANT STEP IN PLANT GENETIC ENGINEERING

Under appropriate culture conditions, plant cells can multiply and form organs such as roots, shoots, embryos, leaf primordia, and can even regenerate a whole plant. The production of GE plants requires regeneration of a whole plant from tissue-cultured plant cells. Using plant tissue cultures, whole plants can then be produced bearing the introduced genetic trait by manipulating single cells in culture. Cultured plant cells can also be used for the mass production of clones, which are genetically identical plants with desired traits. For instance, this approach of clonal propagation using plant tissue culture is commonly used in many ornamental plants commercially.

A major outcome of the plant genome projects is the use of newly identified genes for crop enhancement. Specific genes can be introduced into plants, yielding transgenic lines, exhibiting desirable characteristics. The process cloning is simpler in plants compared to that in animals. In plants, many somatic (not germ line) cells are totipotent, and can develop into whole plants under suitable conditions such as proper plant material to start with, appropriate nutrients, and timing and concentration of hormonal treatments to maximize growth differentiation. A whole plant can be regenerated from a small tissue or plant cells in a suitable culture medium under controlled environment (Fig. 4.15). The plantlets so produced are called tissue-culture raised plants. These plantlets are a true copy of the mother plant and show characteristics identical to the mother plant.

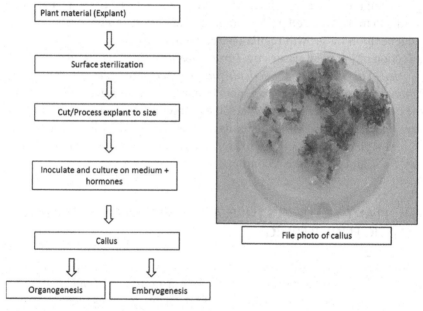

FIGURE 4.15   Flowchart of plant tissue culture.

With the help of tissue culture and transformation techniques, desirable genes can be introduced into plants, yielding transgenic cells and tissues. Whole plants can then be regenerated using tissue culture. The process is much simpler in plants than in animals. Many somatic plant cells are *totipotent*, expressing portions of their previously unexpressed genes to finally develop into whole plant under the right conditions. Most plant tissue cultures are initiated from explants, or small tissue sections from an intact plant which have been removed under sterile conditions. After being placed on a sterile growth medium containing nutrients, vitamins, and combinations of plant growth regulators, cells present in the explant will begin to divide and proliferate.

Depending on the type of plant tissue used as the explant (the starting material) and the composition of the growth medium, plant tissue cultures can be broadly grouped as follows:

4.4.1 CALLUS CULTURE

An explant, usually containing meristematic cells, is incubated on the growth medium containing plant growth regulators such as auxin and cytokinin. The cells that grow from the explant divide to form an undifferentiated mass of cells called a callus. Cells can proliferate indefinitely if they are periodically transferred to fresh growth media and can be directed to differentiate into roots and/or shoots (organogenesis) if they are transferred to a growth medium containing different combinations of plant growth regulators.

4.4.2 PLANT CELL SUSPENSION CULTURE

Plant cell suspension culture involves the transfer of plant callus cells into liquid medium, containing plant growth regulators in specific concentration and chemicals that promote the cells to disaggregate into single cells or smaller clumps of cells in a continuous shake culture. Suspension cultures are also convenient for producing and collecting the plant chemicals that the cells secrete, thus reducing the processing time and expenses. In addition, through somatic cell embryogenesis, plant suspension cell cultures can be used to produce whole plants where the medium contains a combination of growth regulators that drive differentiation and organization of the cells to form individual embryos.

4.4.3 PROTOPLAST ISOLATION AND CULTURE

Protoplasts are plant cells whose thick cell walls have been removed by an enzymatic process, leaving behind a plant cell enclosed only by the plasma membrane. This process has also been useful because plant protoplasts usually begin to resynthesize cell walls within hours of their isolation. Plant protoplasts are also easily transformed with foreign DNA by electroporation method.[38] Thus, protoplast fusion provides an additional means of genetic engineering, allowing traits from one plant to be incorporated into another plant despite natural differences between the species. When either single or fused protoplasts are transferred to specific growth medium, cell wall regeneration takes place, followed by cell division to form a callus. Once a callus is formed, whole plants can be produced either by organogenesis or by somatic cell embryogenesis in culture.

4.4.4 ANTHER/POLLEN CULTURE

The anthers are parts of a flower that contain the pollen and allow their dispersal for normal flower development. For anther culture, anthers are excised from the flowers of a plant and transferred to an appropriate growth medium. The pollen cells from anther can be manipulated to form individual plantlets, which can be cultured and used to produce mature plants through the formation of embryos. These plants are sterile and are usually haploid because they are originally derived from pollen cells that have undergone meiosis. However, these plants can be treated at an early stage with chemical agents such as colchicine resulting in fertile diploid organisms, which are homozygous for every single trait, dominant or recessive. The homozygous plants are useful tools for breeders to introduce a naturally recessive trait.

4.4.5 CULTURE OF PLANT PARTS

Plant organs can also be grown under culture conditions, and provides a useful tool in the study of plant organ development. For example, pollinated flowers of a plant can be excised and transferred to a culture flask containing an appropriate nutrient medium. Over time, the ovular portion of the plant develops into a fruit. Sections of plant roots can also be excised and transferred to a liquid growth medium in which, the roots can proliferate extensively, forming both primary and secondary root branches.

4.5   WORLD POPULATION IN RELATION TO ADVANCES IN CROP PRODUCTION

The genetic engineering of plants plays a major role in resolving the problem of food scarcity of an increasing world population. Biotechnology and improved crop practices are now being employed to transform plants with improved resistance to disease, insects, herbicides, and viruses and nutrient quality of seed grains. The plants can be engineered to tolerate and thrive in stress such as heat or salt. Compared with traditional methods that rely on plant breeding, genetic engineering can reduce the time frame required for the development of crop varieties with better productivity and resistance to biotic and abiotic stress. Moreover, the genetic barriers, such as pollen compatibility with the pistil, no longer limit the introduction of advantageous traits through genetic engineering. A useful trait can be incorporated into a crop plant, once it is identified at the level of genes, by introducing the DNA bearing these genes into the crop plant genome.

4.5.1   USEFUL TRAITS THAT CAN BE INTRODUCED INTO PLANTS

Foreign DNA can be incorporated into an existing plant genome by plant transformation. At present, there are several approaches for plant transformation, the most frequently used method is *A. tumefaciens* mediated.[34] Plant transformation represents a technology which being used extensively for phytoremediation, to improve nutritional quality of foods, and produce "edible plant vaccines."

4.5.2   IMPROVED NUTRITIONAL QUALITY OF FOOD CROPS

Approximately 75% of the world's production of oils and fats come from plant sources. For medical and dietary reasons, the use of high-quality vegetable oils is promoted. Genetic engineering has allowed researchers to produce "designer oils" for both nonedible and edible products.[39] In one technique, canola oil has been modified to replace cocoa butter as a source of saturated fatty acids; enzyme ACP desaturase has also been modified for making monounsaturated fatty acids in transgenic plants. The amino acid contents of various plant seeds are being modified to present a more complete nutritional diet to the consumer. Livestock feed require lysine supplementation, which can be addressed by using a high-lysine corn seed.[40] Fruits and vegetables, such as tomatoes, may be engineered to contain increased levels of nutrients such as betacarotene, and vitamins A and C, which may help to fight and protect against chronic diseases.

4.5.3   PHYTOREMEDIATION

Cleaning up environmental toxins to reclaim polluted land is a challenge. Plants can be genetically modified to accumulate heavy metals that are present in the environment at high concentrations. Because most of their biomass of these plants is water, the dried plants allow for the collection of the metals in a small area. Organic compounds that pose hazards to human health can to be taken up by plants and broken down into harmless components. A branch of biotechnology, "metabolic engineering" modifies biochemical pathways and is also being used to break down toxic substances. For instance, poplars have been engineered to break down TNT.[41]

4.5.4   PLANTS BEARING THERAPEUTICS FOR HUMAN DISEASES

The introduction of "vaccine coding genes" into edible plants is another very interesting application of plant genetic engineering. The genes encoding the antigen (for example, a viral protein) for a particular human pathogen are introduced into the genome of an edible plant such as a banana, tomato, or apple via plant transformation. This antigen protein would then be produced in the cells of the edible part of the plant, and an individual who consumes it would develop antibodies against that pathogenic organism. Edible vaccines are being developed for a coat protein of hepatitis B, an enterotoxin B of *E. coli*, and a viral capsid protein of the Norwalk virus[42] (see Section 4.6, this chapter). This is of great advantage in tropical areas where it is difficult to maintain low temperatures to keep the traditional vaccines (which are generally proteins) cold.

Edible vaccine production in potato: The bacterium, *A. tumefaciens* is commonly used to deliver the genetic material encoding bacterial or viral antigens. The bacterial cells (*A. tumefaciens*) are transformed with a plasmid that carries the gene encoding an antigen and an antibiotic resistance gene. The potato leaves which have been cut into pieces are exposed to the antibiotic which selectively kills the untransformed cells (Fig. 4.16). The surviving transformed cells form a callus. This callus is grown on selectively supplemented medium to sprout shoots and roots, which are grown in soil to form plants. In nearly 3 weeks, the plants bear potatoes which express antigen vaccines (Fig. 4.16).

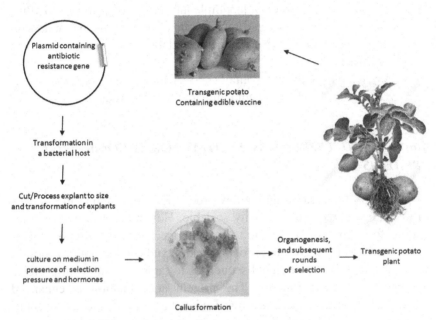

FIGURE 4.16 Edible vaccine production in potato.

4.6 MOLECULAR FARMING

Although plants have been used to produce grains, fiber and several important molecules, they can be GE (plants) to produce desired molecules on a large scale, a process also known as "molecular farming/pharming." The plants are GE mainly by integrating transgenes into the nuclear genome, and screening for maximum expression of one or a few transgenes. This technique could be used to meet the increasing demand for modern medicines, and cater to the healthcare, especially in developing and poor countries. Molecular farming in plants was attempted for the first time in 1989 with the production and functional recombinant antibodies could be expressed in tobacco.

Advantages of using plants for the purpose of protein production include[43, 44]:

- *Cost reduction*: significantly lower production costs than with transgenic animals, mammalian cell or microbial cultures because no requirement of fermentors or bioreactors; plants can be grown on greenhouses or, even in simple lab setups to promote growth;
- *Stability*: plant cells can direct proteins to environments, or subcellular compartments that reduce degradation and therefore increase stability;
- *Safety*: plants do not contain known human pathogens (such as virions, etc.) that could contaminate the final product.

4.6.1 PLANT EXPRESSION SYSTEMS FOR RECOMBINANT PROTEINS

Plants are useful because they allow production of protein at high-levels in just a few days, and are easily scalable. However, for a plant homozygous for a specific transgene to be generated, it takes several generations. Methods for increasing protein yield include codon optimization of the genes, and fusion of signal sequences to target recombinant proteins to subcellular compartments. The protein accumulation is enhanced when the signal sequences are fused with the gene of interest, in addition to providing protection from degradation by enzymes in the host cell. Various viral vectors have been developed for small- or medium-scale plant molecular farming (PMF) products (Fig. 4.17). For example, scientists could develop a highly efficient, bean yellow dwarf virus (BeYDV)-based single-vector DNA replicon system, which incorporated multiple DNA replicon cassettes to produce antibody in tobacco leaves within 4 days following infiltration.[113]

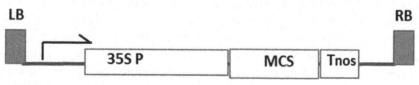

FIGURE 4.17   Diagram of a vector with 35s promoter cassette.

4.6.1.1   PLANT CELL-CULTURE SYSTEM

The plant cells are cultured in a sealed, sterilized container system, similar to microbial or mammalian cell bioreactors, taking care to eliminate any human pathogens or soil contaminants. However, the operational cost is much less expensive than mammalian or microbial bioreactors because cultured plant cells require only simple nutrients to grow. This also removes the biosafety concerns associated with the distribution of pollen, which is unintended and also puts an end to cross-fertilization. Due to the absence of plant fibers and secondary metabolites, downstream purification and processing of the recombinant protein is simplified, thus significantly reducing production costs. Moreover, the downstream purification process can be made further less complicated when the recombinant protein is expressed to be secreted into the culture medium. However, some of the large sized proteins may be checked due to the pore size in plant cell.

4.6.1.2   ALGAE CULTURE SYSTEM

Microalgae have a very simple structure, and can be unicellular, colonial, or filamentous. Because of their short life cycle, algae can produce large amount of biomass within a very short period. The downstream purification of recombinant proteins in algae is generally less expensive than whole plant production systems, being similar to yeast and bacterial systems.[45] However, the algal system may not be suitable for the production of some glycoproteins because recombinant proteins produced from algae do not undergo certain posttranslational modifications. Several diagnostic and therapeutic recombinant proteins, including vaccines, enzymes, and antibodies, have been produced in algal systems.

4.6.1.3   OTHER PLANT PRODUCTION SYSTEMS

Currently, most pharmaceutical proteins are synthesized in leafy crops for optimum biomass. Leaf proteins, however, are subject to rapid proteolytic degradation after they are long-term storage of leaf material is also very challenging. Foreign protein overexpression in leaf cells may also result in necrosis and finally, cell death. Transient expression of various blood clot-dissolving serine proteinases, such as vampire bat plasminogen activator

(DSPAα1), nattokinase, and lumbrokinase, in leaves have resulted in leaf necrosis in just 4 days after infiltration. *Seed-based systems* for PMF have been developed in various plant species, including *Arabidopsis*, tobacco, rice, and corn. Very high levels of recombinant proteins accumulation have been reported when targeted to seeds. A relatively protective environment is present in the *endoplasmic reticulum* (ER) *compartment* of plant cells because it has been sown to contain few proteases. Thus, in the absence of protein degradation and protein stability, the yield increases manifold in the ER. KDEL, an ER signal peptide, was used to target the deposition of a recombinant protein to the ER.[46]

4.6.1.4   CHLOROPLAST TRANSFORMATION

The three important requisites of plastid transformation are:

1. A method for DNA delivery through the double membrane of the plastid: The most widely used method for chloroplast transformation is by biolistic bombardment (Fig. 4.18) of the cells with DNA-coated tungsten particles. A stable, alternate transformation in tobacco has also been reported by the polyethylene glycol (PEG) treatment of leaf protoplasts in the presence of plasmid DNA.
2. An efficient selection method for the transformed plastid (transplastome): transformed copies of the genome can be screened for (a) streptomycin resistance encoded by mutant 16S rRNA gene (*rrn16*)[47] that confer spectinomycin and streptomycin resistance; (b) kanamycin resistance encoded by *aphA-6*.[48]
3. Successful integration of the heterologous DNA so that the normal function of the plastid genome is not disturbed.[49]

Advantages: Plastid transformation has several advantages that latter method lacks. There are several advantages of chloroplast transformation over nuclear transformation.

In many particular species, all plastid types carry identical, multiple copies of the same genome. Insertion of foreign gene into plastid genome may result in amplification of 50–100 copies of the gene per cell, thus increasing the productivity.

As plastid genes are inherited in (almost all) crop plant by the female parent only, there is no damage of the introduced gene getting leaked into

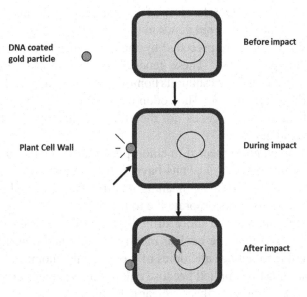

FIGURE 4.18   Biolistic plant transformation.

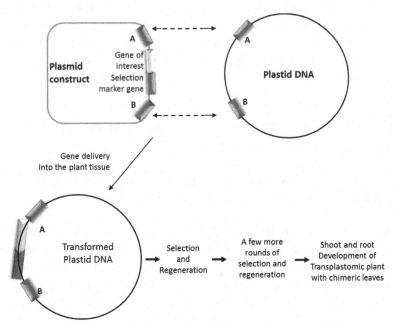

FIGURE 4.19   Transformation of a chloroplast with a recombinant construct. A and B: homologous regions.

wild relatives. Therefore, relocation of nuclear genes to the plastid genome will confine to the transferred genes to the crop (Fig. 4.20).

Chloroplast genes are preceded by the −35 and −10 elements typical of prokaryotic promoters.[50] These genes transcribed by RNA polymerase containing plastid-encoded subunits homologous to the α, β, and β subunits of *E. coli* RNA polymerase. The codon usage of chloroplast genes which are close to prokaryotic genes are therefore a suitable place to express useful bacterial genes.

Transgenic plants subjected to chloroplast transformation are selected after several generations of plants have been regenerated from the gene gun-bombarded leaf explants, meaning that the plant chloroplast genome has had opportunity to incorporate the transgene.

Challenges: It may be more difficult for DNA to cross the plastid double membrane than the nuclear membrane.[49] In addition, for a transformed genome to replace all copies of the original genome, strong selection pressure must be applied because chloroplast genomes are present in much higher copy number than nuclear genomes.

The transgenic plastid genomes are products of a multiple step process, involving DNA recombination, copy correction, and sorting out of plastid DNA copies, or in other words, a complex process. Chloroplast genome can become somewhat unstable following transformation and that gene amplification represents a highly specialized phenomenon that is not easily manipulated.[51]

4.6.2 FACTORS AFFECTING RECOMBINANT PROTEIN PRODUCTION IN PLANTS

Important factors that drive the production of a heterologous protein in plants are:

(a) Genetically transformed plants are extremely susceptible to the effects of environmental factors immediately after being transferring to natural soil. Abiotic stress such as light, drought, salinity, nutritional deficits, and cold has negative effects on plant products.

(b) Demand, commercial value, and cost-effectiveness in comparison with other alternate production systems.

(c) The ability of suitable host crop species to get transformed is another factor that is considered. Transformation is either

Agrobacterium-mediated, or by electroporation, microinjection, and chemical stimulation of protoplasts.

(d) The tissue specificity of the heterologous protein accumulation, which may depend on the use of specific promoters that drive expression in desired tissues (Fig. 4.17). For example, the promoters could be organ-specific for seed, root, leaf, or fruit, etc., constitutive or inducible, strong or weak (Table 4.2).

(e) The storage, harvesting, and downstream processing of the heterologous protein.

(f) A transgene could be inactivated by methylation that play an important role in endogenous gene expression.

(g) Importantly, codon optimization also enhances, to a certain extent, protein expression.[52]

(h) Posttranslational modification processes and more importantly, stability of the recombinant protein, which in turn, depend on the abovementioned factors.

4.6.3 ISOLATION AND PURIFICATION RECOMBINANT PROTEIN FROM PLANTS

The isolation and purification of the heterologous protein may be greatly facilitated by sequestering the protein into a particular cellular compartment. They may undergo specialized folding and posttranslational modification that requires components of the ER. If an appropriate fusion or signal peptide sequence responsible for directing expression and deposition is included, the recombinant proteins can be targeted to the lumen of the ER, vacuole, or other cellular compartments.

Secretion may facilitate proper folding, and thus has also been found to enhance protein stability. Targeting signals can be used to intentionally retain recombinant proteins within distinct compartments of the cell to protect them from proteolytic degradation, preserve their integrity and to increase their accumulation levels.[53] In this direction, it is now possible to design gene constructs which contain ER-targeting signal peptide, KDEL, and to increase the level of accumulation of foreign proteins in transgenic plants. For example, sequence coding for this tetrapeptide was added to the the gene for the pea seed protein vicilin. In lucerne and tobacco leaves, the level of vicilin-KDEL protein was 20 and 100 times higher than that of the unmodified vicilin, respectively.[54]

The protein synthesis pathway is highly conserved between plants and animals, some important differences in posttranslational modification. The main difference between proteins that are produced in animals and plants, however, concerns the synthesis of glycan side chains. These minor differences in glycan structure could potentially change the bio-distribution, activity, and half-life of recombinant proteins compared with the native forms. The plant-specific glycans can possibly induce allergic responses in humans.

4.6.4   HOST PLANTS FOR PRODUCTION OF RECOMBINANT PROTEINS

A few factors that need to be considered while choosing a particular plant system are the yield of functional protein in a given species against the total biomass yield over a given planted area, any associated overhead costs, and the storage and distribution of the product. For example, the costs of storage and distribution of grain are minimal compared with those of freshly harvested tobacco leaves or tomato fruits, but the costs of extraction and purification are higher from seeds than from softer plant material. Most plant production systems can be broadly classified into following groups.[55]

4.6.4.1   TOBACCO AS A FACTORY FOR RECOMBINANT PROTEIN PRODUCTION

The first recombinant protein from plants was human serum albumin, initially produced in 1990 in transgenic tobacco and potato plants. A well-developed technology for gene transfer and expression, a high biomass yield, a rapid scale-up potential because of good seed production are a few advantages of using tobacco as a production system. Tobacco is a very good candidate for PMF production since it is not a food crop and cannot contaminate other crops by the spread of transgenic pollen.[44,56] The procedure for gene transfer and expression in tobacco is also simple and well established. Transgenic tobacco plant can be grown in as less as 6 months to produce the protein of interest in both seeds and leaves. Some of the proteins such as cholera toxin B subunit, Human growth hormone, a tetanus toxin fragment, and serum albumin have been produced at high levels in tobacco. These proteins have also been found to be structurally authentic and biologically active. However, lack of glycosylation is a major disadvantage of the chloroplast transgenic system. Therefore, chloroplasts are unlikely to be used to synthesize human glycoproteins in which the glycan chain structure is crucial for protein activity.

4.6.4.2 CASE STUDY

Tobacco plants produce cancer vaccine: A therapeutic vaccine protein produced by GE tobacco plants has been effective in laboratory mice in preventing the growth of non-Hodgkin's lymphoma cells.[57] The most prevalent form of lymphoma, Non-Hodgkin's B-cell lymphoma affects the lymph system. The researchers removed malignant B cells from laboratory mice, and isolated the gene coding for their specific surface markers. They inserted this gene into a tobacco mosaic virus (TMV), and exposed tobacco plants to the modified virus. As the virus infection spread through the leaves, the desired B-cell protein was produced. This protein was extracted and injected into mice which had received lethal dosages of tumor cells. Thus, the plant-based expression system has proven itself to be faster than other technologies for producing effective vaccines.

Below mentioned are a few other recombinant biopharmaceuticals which have been produced in tobacco are[56] (the list is not exhaustive):

- H5N1 vaccine (HAI-05)
- HIV P2G12 antibody: through stable nuclear expression
- anthrax recombinant protective antigen vaccine
- malaria vaccine Pfs25 VLP
- Ebola virus ZMApp; non-Hodgkin's lymphoma vaccine; influenza vaccine: through transient expression
- Fabry disease PRX-102: in tobacco cell culture
- anthrax PBI-220, therapeutic recombinant protein
- rabies vaccine
- rotavirus vaccine
- growth hormone
- human serum albumin
- human secreted serum alkaline phosphatase: secretion from roots and leaves
- hepatitis B virus envelop protein

- erythropoietin: produced in tobacco suspension cells
- diabetes autoantigen
- cholera toxin B subunit: expressed in chloroplasts

4.6.4.3 RECOMBINANT PROTEIN PRODUCTION: CEREALS AND LEGUMES

A few advantages of using cereals for recombinant protein production include long-term storage at ambient temperatures, appropriate biochemical environment essential for protein accumulation, and reduced exposure of stored proteins in seeds to nonenzymatic hydrolysis and protease degradation. In addition, lack of phenolic compounds in cereals compared to that in tobacco leaves is also an added advantage. A few cereal crops that are being used for commercial production crop for recombinant proteins are: *maize*, for the production of enzymes such as laccase, trypsin, and aprotinin, and recombinant antibodies[43]; *barley* as bioreactors for lysozyme, thermo-tolerant hybrid cellulase, α1- antitrypsin, human antithrombin III, serum albumin, lactoferrin, and human vascular endothelial growth factor (VEGF) is even being used for treatment for thinning hair; *rice* for production of human lactoferrin. In addition, although the biomass produced by *soybean* and *alfalfa* is lower than tobacco, they carry out nitrogen fixation, thus being more environment-friendly.[56]

4.6.4.4 RECOMBINANT PROTEIN PRODUCTION: VEGETABLES AND FRUITS

The fruits, vegetables, and leafy salad crops can be consumed raw or partially processed. Moreover, the plants expressing these vaccines may be grown locally, where needed most, no transportation costs, naturally stored. *Potatoes* are being widely used for producing plant-derived vaccines, bulk-production system for antibodies,[55] diagnostic antibody-fusion proteins, and human milk proteins. The first plant-derived rabies vaccine was produced in *tomatoes*,[55] which are more palatable and offer high biomass yields than potatoes. Edible recombinant vaccines are also produced in *lettuce*, for example, a vaccine against HBV.[56] Recombinant vaccines produced in *bananas* can be consumed raw or as a puree by both adults and children.[56] Examples of other edible vaccines include pig vaccine in corn, HIV-suppressing protein in spinach, and human vaccine for hepatitis B in potato (Table 4.2).

TABLE 4.2 Listed Below Are Examples of a Few Recombinant Proteins Produced in Plants (the List Is Not Exhaustive) by Molecular Farming.[53–57]

Plant system	Product
Tobacco leaves	α-amylase from *Bacillus licheniformis*
Turnip	α-interferon
Maize; maize seed	Aprotinin; *Acidothermus cellulolyticus* cellulosic degrading enzymes; cellobiohydrolase of *Trichoderma reesei*
Carrot cell culture	Recombinant glucocerebrosidase for treating Gaucher disease
Algae	Enzymes, vaccines and growth factors
Rice	Serum albumin, human Interleukin-10, human lysozyme, human lactoferrin, transglutaminase, α-1 antitrypsin,
Safflower	Apolipoprotein, insulin
Duckweed	Thrombolytic drug
Lettuce, potato	Vaccine against HBV
Tobacco	Antibodies against Ebola and hepatitis B viruses, anti-PA mAb, immunoglobulin A, Fv antibodies
Arabidopsis	IgA antibodies for passive immunization against enterotoxigenic *Escherichia coli*; human intrinsic factor for vitamin B12 deficiency
Chlamydomonas	LSC (HSV)
Potato	*E. coli* heat labile toxin B subunit, *Vibrio cholerae* vaccine, Capsid protein Norwalk virus
Alfalfa	Plasma protein
Brassica seeds	Hirudin
Spinach	Antigen vaccine against *Bacillus anthracis*

4.6.5 MOLECULAR FARMING: CONCERNS AND CHALLENGES

There is no precise and standardized control over the expression of these transgenes, making it difficult to specify a particular dosage of edible vaccines for kids and adults. An important health safety concern is the side effects, nontarget organ responses, and immune response that could be triggered by intake of these plant-based vaccines. Finally, there are also chances of degradation of these vaccines by the digestive enzymes. There are also chances of accident or unintentional contamination of the food supply. Moreover, while purifying a recombinant product from a plant at an industrial level, researchers face difficulty in removing plant pathogens, phenolic compounds, secondary metabolites, pesticides, and fertilizers. While the overall strategy used for downstream processing has to be robust, economically viable, and cGMP compliant, current downstream processing and purification of PMF products are tedious and costly. The process also raises biosafety concerns, such as environment contamination, gene flow via pollen to nontransgenic crops, etc.; the risk is that transgenes and/or their encoded proteins could spread in the environment and that nontarget organisms could be affected.

Therefore, both standard biosafety procedures and purification protocols need to be established. One needs to address several questions, including:

a. Are PMPs safe and effective medicines for humans and animals? It is difficult to assess the impact of such genes on the survival of wild types of the transgenics, and their effects on the natural ecosystems. For this, the regulatory structures are adequately strengthened to regulate and monitor biopharm crops, including full disclosure of field trials, crop, gene, and location.
b. Is it really possible to reduce the production costs of PMPs so that the expected economic benefits are attained both for the crop-based pharmaceuticals and the farmers.
c. To provide acceptable levels of gene containment by working out a proper combination of species, growth environment, and safety procedures. This would prevent the possible spread of selectable marker genes and vector backbone fragments, and the risk of herbicide resistance markers spreading to weeds.

4.7 PLANT STRESS RESPONSES

4.7.1 ABIOTIC AND BIOTIC STRESS

Plants, similar to other living beings have to deal with various environmental factors and have evolved mechanisms to help them adapt to and survive in stressful environments. Exposure of plants to biotic and abiotic stress leads to changes in plant physiology, ultimately leading to a reduction in productivity. Abiotic stress has a huge impact on growth which can result in as high as >50% productivity in most plant species. Biotic stress involves attack on the plants by pathogens and herbivores. The plant defense to these stresses includes timely perception and efficient response. The defense mechanisms of the plant include activation of complex signaling cascades which may vary from one stress to another, including activation of specific ion channels and kinase cascades,[58,59] accumulation of reactive oxygen species (ROS),[60] phytohormones such as abscisic acid (ABA), salicylic acid (SA), jasmonic acid (JA), and ethylene (ET),[61] and a reprogramming of the genetic machinery. These defense reactions enable the plant to tolerance stress so that the biological damage caused is minimized (Fig. 4.20).

Salinity is another restricting factor of PMF. Approximately one-third of the world's irrigated farms are ineffective due to the excess salt content of the soil.[62] The adverse effects of salt on plants are manifested in two ways. First, a high concentration of salt in the soil directly hampers water absorption by the roots by affecting root–soil osmotic regulation. Second, salt accumulation in various organs poisons plants.[63] The two toxic ions derived from NaCl, Na^+, and Cl^- can damage plant cells through both osmotic and ionic mechanisms.[64] Quantitative and qualitative changes in metabolite synthesis as well as the occurrence of enhanced metabolic toxicity are a few of the most usual indicators of stressed plants.[65] Furthermore, salt stress alters the expression of cell cycle progression genes through affecting mitotic cell division.[66] It has been well documented that the ABA content of plants increases under salt-stress conditions.[67] Adaptation to saline stress is accompanied by alterations in the level of numerous metabolites, proteins, and mRNAs. A variety of genes, the expression of which is activated in response to salt stress, have been identified and have been transferred to plants.[68] Because high salinity conditions promote plant cell dehydration,[69] many of the genes that are activated by saline stress are also activated by drought. The expression of the majority of these genes is regulated by ABA, a plant hormone that is generated in response to saline stress.

4.7.2 STRESS RESPONSES AND ROS

There is a rapid generation of ROS in plants after sensing stress.[70] One of the major roles of ROS is to serve as signaling molecules in the cells.[71]

Although ROS have long been known to be destructive and harmful compounds in a stressed organism, their production is fine-modulated by the plant to avoid tissue damage.[72,73] This again might attenuate the oxidative stress caused by abiotic stress.[72] Furthermore, ROS could interfere in *cross-tolerance*, a phenomenon in which some plants are more susceptible when confronted with each stress individually compared to a simultaneous exposure to two different stresses. In addition, certain environmental stresses predispose the plant to respond faster and with better resistance.[74] ROS has been observed in stress-induced tolerance in *A. thaliana* after infection with the pathogen *Verticillium* spp. that infects the vascular tissues by de novo xylem formation resulting in enhanced water flow, thus increasing drought tolerance.[75]

Although ROS are produced in both abiotic and biotic stress, their response differs from one stress to another. ROS production can be sensed by ROS-sensitive transcription factors (TFs) leading to the induction of genes participating in the stress responses. ROS induce stress tolerance by activating response factors such as TFs, mitogen-activated protein kinases (MAPKs), antioxidant enzymes, and pathogenesis-related, temperature-induced-, heat shock-proteins.[76]

MAPKs are responsible for the signal transduction of various cellular processes in response to various abiotic and biotic stress, and certain kinases are involved in both kind of stress.[77] For instance, in cotton, the kinase GhMPK6a negatively regulates both biotic and abiotic stress,[78] such as pathogen attack, exposure to UV radiations.

4.7.3   HORMONE SIGNALING UNDER STRESS

Hormones allow defense responses against defined environmental conditions. The hormone ABA is considered involved in the perception of many abiotic stresses. While the abiotic stress-regulation network is mediated by increases in ABA concentration,[72,79,80] the biotic stress responses are mediated mainly by antagonism between other hormones such as SA and JA/ET. In certain cases, following infection, ABA has been shown to accumulate. However, under the combination of abiotic and biotic stress, ABA mostly acts as opposed to SA/JA/ET inducing a susceptibility of the plant against disease and herbivore attack. [72,79,80] However, since the stomatal closure is another secondary effect of abiotic stress, the entry of biotic stress inducers through these ports of entry in the plant is prevented. Therefore, in such conditions, the plant is protected from biotic as well as from abiotic stress (Fig. 4.20).

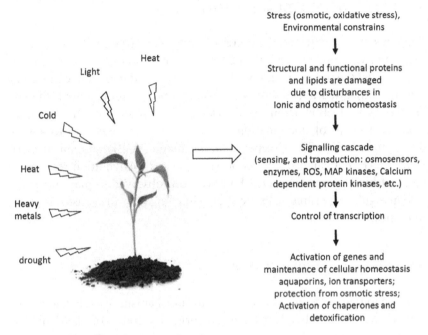

FIGURE 4.20   Effect of stress on plants.

4.7.4   TFs AND MOLECULAR RESPONSES IN CROSS-TOLERANCE

The detection of a stress leads to alterations in gene expression as a result of reprogramming of the molecular machinery regulated by and the TFs.[81,82] Functional genomics have discovered various gene families, which ensure higher production and adjustment to abiotic stresses.

These groups of genes can be expressed ectopically, or delivered to the crops that lacking these genes.[49] Specific phytohormones such as ABA, SA, JA, and ET play very important role in regulating gene activities involved in plant defense. For example, in *Arabidopsis*, TF MYB96 plays an important role in plant protection under pathogen infection by mediating the molecular link between drought stress-induced ABA and pathogen infection-stimulated SA expression.

4.7.5   PLANT RESISTANCE PROTEINS

Plant resistance (PR) proteins are crucial for resistance against pathogens, and when plants are attacked, their expression is strongly upregulated. When the plants are exposed to cold stress and infection, certain TFs are overexpressed and cold-responsive PR genes are activated, thereby conferring protection against both stresses. PR proteins are also accumulated as a result of upregulation of some TFs after exposure to abiotic stress. In addition, TFs C-repeat binding factors (CBF), no apical meristem ATAF, cup-shaped cotyledon (NAC), and Dehydration-responsive element-binding proteins (DREB) have been observed to play important role in regulating primary abiotic stress signaling pathways ensuring stress tolerance.[79,80,83–85,86]

4.8   RNA INTERFERENCE IN PLANTS

The phenomenon of gene silencing-related mechanisms was first noted as a surprise observation by plant scientists during the course of plant transformation experiments. The introduction of a transgene into the plant genome led to the silencing of both the transgene and homologous endogenes.

4.8.1   RNAi AND BIOTIC STRESS IN PLANTS

The RNAi mechanism has been discussed in Figure 4.21. RNAi has also been exploited in plants for resistance against pathogens such as pests, viruses, and nematodes that cause significant economic losses. Keeping aside the significance in growth and development and the maintenance of genome integrity, RNAi-induced gene regulation is vital in plant stress management. Hence, investigating the role of small RNAs in regulating gene expression assists the researchers to explore the potentiality of RNA silencing in abiotic and biotic stress management. This novel approach opens new avenues for crop improvement by developing disease-resistant, abiotic or biotic stress tolerant, and high yielding elite varieties.[87]

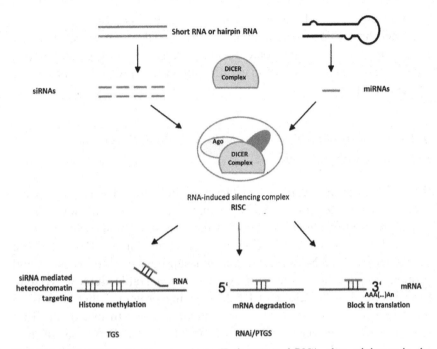

FIGURE 4.21 RNA silencing in plants. Endogenous dsRNA, that originates in the organism, initiates the RNA silencing process by first activating the Dicer protein, a nuclease which interacts with other proteins, such as AGO-2, through its PAZ domain. The dsRNA is cleaved into small interfering RNAs (siRNAs) by the Dicer into two double-stranded fragments with 20–25 base pairs and a 2-nucleotide overhang at the 3' end. The newly formed duplex siRNAs are recruited by the proteins forming a complex RNA-induced silencing complex (RISC). RISC is activated by unwinding of the duplex and displays nuclease activity toward corresponding mRNA. The antisense strand of the siRNA mediates sequence specificity and target recognition.

4.8.1.1   PLANT DISEASE RESISTANCE

Pathogens can cause a huge reduction in crop yield that can lead to significant economic losses. While plant pathologists and biotechnologists have adopted different approaches to develop pathogen-resistant genotypes, in last decade, RNAi-induced gene silencing has emerged as a tool to effectively generate plants that are pathogen-resistant. This approach proved to be effective in creating resistance against bacteria, fungi, and viruses. The precursor in RNA interference is the double-stranded RNA (dsRNA) which activates the RNAi machinery to inhibit the transcription and translation to silence the specific genes.[88] This approach has opened new avenues in the development of eco-friendly techniques for plant improvement and resistance as specific genes which cause stress are suppressed and certain genes are expressed for providing disease resistance.

4.8.1.2   RESISTANCE AGAINST VIRAL DISEASES BY VIRUS-INDUCED GENE SILENCING

RNA silencing also causes resistance against viral diseases by virus-induced gene silencing (VIGS) mechanism.[88,89] RNA silencing hosts target protein translation and process the virus-mediated dsRNA, which results by pathogen replication into virus-mediated siRNAs. These virus-mediated siRNAs then target and suppress gene expression and protein translation of the virus genes. For a stable defense system, viruses encode suppressor of RNA silencing proteins,' that have been identified and isolated from various plant viruses. In one of the first experiments,[90] demonstrated for the first time that PPV coat protein plays a role in RNAi for virus resistance in woody perennial species and produced *Plum pox virus* (PPV)-resistant plants. In addition, proper antisense or hairpin RNAi constructs also exhibit RNAi and confer plant resistance.

4.8.1.3   HOST GENE SILENCING BY HAIRPIN RNAi

The use of short hairpin RNAs (shRNAs) is another form of RNAi which is synthesized within the cell by DNA vector-mediated production. The hairpin RNA consists of two 19–22 bp complementary RNA sequences linked by a short loop of 4–11 nucleotides, which resemble a hairpin structure. This method can be employed to increase disease resistance by changing the gene expression against fungal and bacterial pathogens through genetic engineering of the host plant. For example, bacterial flagellin can regulate specific miRNA expression to increase disease resistance signaling pathway in *Arabidopsis*.[91] In rice, RNAi can knockdown the gene controlling fatty acid desaturase activity and cause increased resistance blast fungus (*Magnaporthe grisea)* and against the bacterial pathogen (*Xanthomonas oryzae* pv. *Oryzae*).[92] Gene silencing can be obtained by *Avra10*, a host-induced gene that results in limited fungal disease attach in wheat and barley. This happens through a transient gene expression which is RNAi-resistant because of point mutations that are silent. Thus, the transfer of RNA from host plant to fungal pathogen *Blumeria graminis* causes RNAi-based plant protection against these pathogens.[93]

4.8.1.4 CROP PROTECTION FROM INSECTS AND PEST RESISTANCE

Although classical breeders have developed various insect/pest-resistant cultivars, this approach is tedious and time-consuming as the added traits increase the complexity. The practice of using pesticides to control pests has become a common approach around the world, but over the years, due to adverse health and environmental effects, its use seems to be very limited. Most of the approaches to control insect/pests on crops are Bt-based. The Bt insecticidal proteins help in the permeabilization of gut epithelial cell's membrane in susceptible insects. Although effective, this approach is limited for some specific crops to manage some specific pests, and some insects can develop resistance against Bt toxin.

Crop protection from insects through genetic engineering to exhibit dsRNAs target for knocking specific pathogen genes has been well documented. RNAi offers a robust and more discriminatory pathway for battling various destructive insect/pests that are economically loss making. In plant-mediated herbivorous insects, RNAi is a strategy in which the suppression of a critical insect-gene through insect feeding on plant engineered to develop a specific dsRNA that can prompt dissection of gene functions in these insects.[94] Herbivorous insect RNAi efficiency can be further be stimulated by ingestion of transgenic dsRNA-producing plants that is gene-specific and proved effective against insect damage. A gene "CYP6AE14" in *Helicoverpa armigera* expresses in the insect midgut and correlates with larva growth when food contains gossypol.[94] Therefore, after feeding on plant material exhibiting dsRNA specific to gene "CYP6AE14," the effect of the transcript decreases in midgut and larva growth also retards.

4.8.1.5 INTERFERENCE IN EXPRESSION OF THE TARGETED GENES

Various phenotype disturbances are caused when expression of the targeted genes are interfered with, namely, stunted growth, moulting defects, and insect mortality. Western corn rootworm (WCR) ingestion of dsRNAs provided in diet can trigger the RNAi which results in stunted larva growth and their death.[95] Further, transgenic corn plants which are engineered for WCR dsRNAs expression exhibit a substantial reduction in insect damage that suggests RNAi pathway is effective and can be exploited further to control coleopteran insects.

4.8.1.6 RNAi IN PLANTS TO DEVELOP RESISTANCE AGAINST NEMATODES

This is a tool to control plant parasitic nematodes.[96] dsRNAs can be produced through engineered plants that have the ability to silence target genes in nematode body. The dsRNAs from plant is delivered to nematode by the ingestion of plant cytoplasm and after its entry into the nematode body, accelerate the RNAi to finally result in targeted genes inactivation through dsRNA. To target specific host genes, efficient regulation of dsRNA triggers expression is a prerequisite that will reduce the negative effects on plant growth and development and it also necessitates identification of nematode-responsive promoters. In vitro ingestion of 16D10 dsRNA gene is reported to the target parasitism gene silencing in root-knot nematodes and reduced nematode lethality, whereas in *Arabidopsis, in vivo* expression of 16D10 dsRNA gene also increase the resistance against four species of root-knot nematodes.[97] Since, there is no single natural-resistant gene known against root knot nematodes, expression of dsRNA to silence target genes for disruption of the parasitism effect significantly proved a sustainable approach for robust and resistant cultivars. RNAi enables parasitism related to the molecular determinants and also may potentially identify novel specified targets vital for survival of nematodes and thus, leads to the development of efficient control methods.

Plants armed with dsRNA prevent the insect/pest damages, as the transgenic cultivars that produced dsRNAs can target certain genes in insect tissues to reveal dominance in gene expression and caused their mortality.[87] Robust RNAi pathway effective against different insects and pests have open new pathways for crop protection by developing insect-resistant cultivars of commercially important plants. Nevertheless, the success of the RNAi approach to control notorious insect/pests is mainly corelated with the wise screening system for target gene selection and an appropriate delivery mechanism.

4.9   RNAi AND ABIOTIC STRESSES

RNAi plays an imperative role in abiotic stress response stimulation in different crops. The function of microRNA (miRNAs) in relation to abiotic stress such as oxidative stress, cold, drought, and salinity were reported[98] in *Arabidopsis* plants under various abiotic stress. Reports confirmed miR393 was significantly upregulated when experiencing dehydration, higher salinity levels, cold, and ABA. Additionally, miR402, miR319c, miR397b, and miR389a in *Arabidopsis* were controlled by abiotic stress under varying levels.[99] The ability of RNAi to be specific and sequence-based has been effectively utilized for incorporating desired traits for abiotic stress tolerance properties in various plants species.[99]

4.9.1   DROUGHT STRESS TOLERANCE

Drought is a severe ecological stress on agriculture production around the world. Numerous attempts have been made by plant scientists to increase productivity of crops in even in drought conditions. Plant is drought tolerant when it has the potential to uphold enough water inside the tissues and maintain turgidity when faced with a drought condition. Based on the experiments involving gene expressions, drought-specific allele could be broadly classified into three major groups: (1) genes implicated in signal transduction pathways (STPs) and transcription; (2) genes involved in membrane protection and protein activity; and (3) genes involved in ion uptake and water transport.[100]

In relation to drought responses, miR159 has been reported in triggering the signaling of hormones in *Arabidopsis*.[101] Furthermore, miR169g and miRNA393 genes have been observed in rice crop which were stimulated under drought conditions.[102] The RACK1 transgenic rice was observed with a superior level of tolerance in contrast to nontransgenic rice plants.[103]

In many plants, such as *Arabidopsis, Populus trichocarpa,* and *Oryza sativa,* the miRNA expression profiling performed under drought stress reveals miRNA to be drought-responsive.[104]

Analysis of miRNAs and genome sequencing profiling using a microarray platform executed in drought-studied rice at a various range of growth stages, from tiller formation to inflorescence. The results suggest that 16 miRNAs were remarkably involved in downregulation in response to drought stress.[104]

In contrast, 14 miRNAs have been found in upregulation under drought stress. Few miRNAs gene families, such as miR319, miR896, and miR171, are recognized as both up- and downregulated groups.[105] The miR1447, miR1445, miR171l-n, and miR1446a-e in *Populus* have been identified as drought-responsive.[106] Under drought stress, miR169 was found downregulated in the roots only when studied in *Medicago truncatula,* while miR408 and miR398a,b were highly upregulated both in roots as well as shoots.[107] In recent studies, miRNA expression patterns of drought-tolerant wild emmer wheat in relation to drought stress have been explored using a plant miRNA microarray platform.

4.9.2 SALT STRESS TOLERANCE

It has been estimated that nearly 20% of agricultural land is salt-affected, tremendously decreasing efficiency of the agricultural production due to reduction in available irrigation water quality. Many regulated miRNAs have been reported in salinity-stressed plants. For instance, in *P. vulgaris,* an increment in accumulation of miR159.2 and miRS1 with the addition of NaCl has been observed.[108] The investigation using microarray elucidates the miRNA profile of salinity-tolerant and a salt-sensitive line of maize; the findings indicate that members of miR156, miR164, miR167, and miR396 groups were downregulated, while miR162, miR168, miR395, and miR474 groups were upregulated in saline-stressed maize roots.[109]

4.9.3 COLD AND HEAT STRESS TOLERANCE

The miRNA expression in *Arabidopsis, Populus, and Brachypodium* species exposed to cold stress has been studied and miR169 and miR397 are found upregulated in all aforementioned species. Additionally, many miRNAs are induced in *Arabidopsis* under cold stress; on the other hand, some miRNAs exhibit either transitory or gentle regulation when exposed to cold stress.[104] Under heat stress, miRNA in wheat show variant expression; out of 32 distinguished families of miRNA, 9 miRNAs were identified to be heat responsive. For instant, miR172 was distinctly decreased, while miRNAs (including miR827, miR156, miR169, miR159, miR168, miR160, miR166, and miR393) were upregulated in response to heat stress.[110]

4.9.4 UVB RADIATION STRESS TOLERANCE

miRNAs stimulated by UV-B radiation have also been identified. Few of the families were identified to be commonly upregulated in *Arabidopsis* by UV-B radiation (miR168, miR156, miR167, miR160, miR398, and miR165/166) are also upregulated in *Populus termula*. Moreover, three families (miR393, miR159, and miR169) that were found as a downregulating in *P. termula* were upregulated in *Arabidopsis* implying that some UB-V radiation stress responses could be species-specific.[111]

4.9.5 MECHANICAL STRESS TOLERANCE

Plants face mechanical stress when its parts are exposed to external forces, as wind or gravity. In an investigation of *P. trichcarpa*, Pt-miRNA levels of transcript were compared in compression-stressed or tension-stressed xylem with the nonstressed xylem, whereas miR408 showed upregulation, miR156, miR48, miR162, miR475, miR164, and miR480 were downregulated by compressing and tension. Thus, miRNAs may play an important role in maintaining the mechanical and structural fitness of the plants.[112]

4.10 SUMMARY

Plant genomes are very large in comparison to other eukaryotes, mainly due to a high amount of repetitive DNA. Plant genomes can also be compared with one another by mapping the locations of certain genes or gene traits in various plants by using RFLPs and AFLPs techniques.

The discovery of genomic variations and genes associated with adaptation to climatic changes found in wild relatives of crop plants have been accomplished via whole-genome sequencing. The information could be relevant for implementing breeding approaches to develop crops that can adapt environmentally. The importance of integration of various omics approaches for abiotic stress tolerance in model legume crop, soybean, cannot be undermined. There have been significant advances for abiotic stress tolerance in plants in terms of availability of molecular markers, QTL mapping, genome-wide association studies (GWAS), genomic selection (GS) strategies, and transcriptome profiling. Studies in other omics branches such as proteomics, metabolomics, and their integration with genomics is very important and will play a major role in understanding abiotic stress responses in future.

The TFs play a central role in abiotic stress response and tolerance mechanisms. Other molecular components of the signaling pathways, such as MAPKs, ROS, and lipid-derived pathways have also been implicated in plant adaptation to environmental adversity. The use of TFs function in crosstalk among various abiotic stress responses and are increasingly being done to improve abiotic stress tolerance in different crops. Calcium ions especially play a pivotal role in abiotic stress signaling and several signal transduction cascades in plants. The activity of CBL-interacting protein kinases (CIPKs) is modulated by calcineurin B-like proteins (CBLs) which function as calcium sensors. In the conditions of abiotic stresses, the CBL–CIPK network helps in maintaining proper ion balances. VIGS has also emerged as an efficient RNAi tool for gene function analysis in plants.

Despite the existing comprehensive knowledge in this area of plant molecular biology, many problems still remain unaddressed. With the depletion of natural resources, ever-increasing global population and climate change threat, sustainable and higher crop production is greatly needed. Therefore, several approaches and their integration need to be employed to understand the molecular basis of abiotic stress response and adaptation for the development of stress-tolerant crop varieties.

4.11   QUESTIONS

1. What characteristics of *A. thaliana* and rice make it useful as model systems in genetic studies and for the sequencing of its entire genome?
2. How would functional genomics in plants help in dealing with the world food requirements?

3. What role do the microarrays play in functional genomics?
4. What characteristics of chloroplast make it suitable for producing recombinant proteins? What are the challenges in using chloroplast systems?
5. How can RNAi help in developing plant disease resistance?
6. What are the ethical, environmental, and regulatory issues related to molecular farming?

KEYWORDS

- plant genomes
- plant transformation
- molecular farming
- plant stress responses
- RNA interference
- RNAi and abiotic stresses

REFERENCES

1. Mangelsdorf, P. C. George Harrison Shull. *Genetics* 1955, 40, 1–4.
2. Xiao, J.; Li, J.; Grandillo, S.; Ahn, S. N.; Yuan, L.; Tanksley, S. D.; McCouch, S. R. Identification of Trait-improving Quantitative Trait Loci Alleles from a Wild Rice Relative, *Oryza rufipogon*. *Genetics* **1998**, *150*, 899–909.
3. Mehrotra, S.; Goyal, V. Repetitive Sequences in Plant Nuclear DNA: Types, Distribution, Evolution and Function. *Genomics Proteom. Bioinform.* **2014**, *12*(4), 164–171.
4. Kumar, S.; Banks, T. W.; Cloutier, S. SNP Discovery Through Next-generation Sequencing and Its Applications. *Int. J. Plant Genomics* **2012**, *2012*, Article ID 831460, 15.
5. Melters, D. P.; Bradnam, K. R.; Young, H. A.; Telis, N.; May, M. R.; Ruby, J. G.; … Chan, S. W. Comparative Analysis of Tandem Repeats from Hundreds of Species Reveals Unique Insights into Centromere Evolution. *Genome Biol.* **2013**, *14*(1), R10.
6. McClintock, B. The Origin and Behavior of Mutable Loci in Maize. *Proc. Natl. Acad. Sci. USA.* **1950**, *36*(6), 344–355.
7. Devos, K. M.; Gale, M. D. Comparative Genetics in the Grasses. *Plant Mol. Biol.* **1997**, *35*, 3–15.
8. Wicker, T.; Krattinger, S. G.; Lagudah, E. S.; Komatsuda, T.; Pourkheirandish, M.; Matsumoto, T.; … Keller, B. Analysis of Intraspecies Diversity in Wheat and Barley

Genomes Identifies Breakpoints of Ancient Haplotypes and Provides Insight into the Structure of Diploid and Hexaploid Triticeae Gene Pools. *Plant Physiol.* **2009,** *149*(1), 258–270. http://doi.org/10.1104/pp.108.129734.

9. Moore, G.; Devos, K. M.; Wang, Z.; M. D. Gale. Grasses, Line Up and Form a Circle. *Curr. Biol.* **1995,** *5,* 737–739.

10. Brkljacic, J.; Grotewold, E.; Scholl, R.; Mockler, T.; Garvin, D. F.; Vain, P., et al. Brachypodium as a Model for the Grasses: Today and the Future. *Plant Physiol.* **2011,** *157*(1), 3–13.

11. Griffiths, A. J. F.; Miller, J. H.; Suzuki, D. T., et al. *An Introduction to Genetic Analysis;* 7th ed. W. H. Freeman: New York, 2000.

12. Kurata, N.; Nonomura, K.; Harushima, Y. Rice Genome Organization: The Centromere and Genome Interactions. *Ann. Bot.* **2002,** *90*(4), 427–435. http://doi.org/10.1093/aob/mcf218.

13. Vos, P.; Rogers, R.; Bleeker, M.; Reijans, M.; Van de Lee, T.; Hornes, M.; Fritjers, A.; Pot, J.; Peleman, J.; Kuipe, M.; Zabeau, M. AFLP: A New Technique for DNA Fingerprinting. *Nucleic Acids Res.* **1995,** *23,* 4407–4414.

14. The Arabidopsis Genome Iniative. The *Arabidopsis* Genome Initiative. Analysis of the Genome Sequence of the Flowering Plant *Arabidopsis thaliana. Nature* **2000,** *408,* 796–815.

15. Meinke, D. W.; Cherry, J. M.; Dean, C.; Rounsley, S. D.; Koornneef, M. *Arabidopsis thaliana*: A Model Plant for Genome Analysis. *Science* **1998,** *662,* 679–682.

16. Shimamoto, K.; Kyozuka, J. Rice as a Model for Comparative Genomics of Plants. *Annu. Rev. Plant Biol.* **2002,** *53,* 399–419.

17. Libault, M.; Farmer, A.; Joshi, T.; Takahashi. K.; Langley. R. J.; Franklin, L. D.; He, J.; Xu, D.; May, G.; Stacey, G. An Integrated Transcriptome Atlas of the Crop Model *Glycine max*, and Its Use in Comparative Analyses in Plants. *Plant J.* **2010,** *63*(1), 86–99.

18. Mahesh, H. B.; Shirke, M. D.; Singh, S.; Rajamani, A.; Hittalmani, S.; Wang, G-L.; Gowda, M. Indica Rice Genome Assembly, Annotation and Mining of Blast Disease Resistance Genes. *BMC Genomics* **2016,** *17,* 242.

19. The International Wheat Genome Sequencing Consortium (IWGSC). A Chromosome-based Draft Sequence of the Hexaploid Bread Wheat (*Triticum aestivum*) Genome. *Science* **2014,** *345*(6194), 1251788. DOI:10.1126/*science*.1251788.

20. The Tomato Genome Consortium (TGC). The Tomato Genome Sequence Provides Insights into Fleshy Fruit Evolution. *Nature* **2012,** *485*(7400), 635–641.

21. Nannas, N. J.; Dawe, R. K. Genetic and Genomic Toolbox of *Zea mays. Genetics* **2015,** *199*(3), 655–669.

22. Horton, M. W.; Hancock, A. M.; Huang, Y. S.; Toomajian, C.; Atwell, S.; Auton, A, Muliyati, N. W.; Platt A; Sperone, F. G.; Vilhjlmsson, B. J.; et al. Genome-wide Patterns of Genetic Variation in Worldwide *Arabidopsis Thaliana* Accessions from the Regmap Panel. *Nat. Genet.* **2012,** *44* (2), 212–216.

23. Daniel, H.; Lin, C-S.; Yu, M.; Chang, W.-J. Chloroplast Genomes: Diversity, Evolution, and Applications in Genetic Engineering. *Genome Biol.* **2016,** *17,* 134.

24. Cullis, C. A.; Vorster, B. J.; Van Der Vyver, C.; Kunert, K. J. Transfer of Genetic Material Between the Chloroplast and Nucleus: How Is It Related to Stress in Plants? *Ann. Bot.* 2009, *103*(4), 625–633.

25. Turmel, M.; Otis, C.; Lemieux, C. The Complete Chloroplast DNA Sequence of the Green Alga *Nephroselmis olivacea*: Insights into the Architecture of Ancestral Chloroplast Genomes. *Proc. Natl. Acad. Sci. USA* **1999,** *96*(18), 10248–10253.

26. Shaw, J.; Lickey, E. B.; Schilling, E. E.; Small, R. L. Comparison of Whole Chloroplast Genome Sequences to Choose Noncoding Regions for Phylogenetic Studies in Angiosperms: The Tortoise and the Hare III. *Am. J. Bot.* 2007, *94*(3), 275–88.

27. Jeffrey, D. P.; William, F. P. Chloroplast DNA Rearrangements Are More Frequent When a Large Inverted Repeat Sequence is Lost. *Cell* **1982,** *29*(2), 537–550.

28. The Chloroplast: Interactions with the Environment; Anna Stina, A., Aronsson, H., Eds.; Springer: 2009, pp 18. ISBN 978-3-540-68696-5.

29. Sablok, G.; Padma Raju, G. V.; Mudunuri, S. B.; Prabha, R.; Singh, D. P.; Baev, V.; Yahubyan, G.; Ralph, P. J.; Porta, L. ChloroMitoSSRDB 2.00: More Genomes, More Repeats, Unifying SSRs Search Patterns and on-the-Fly Repeat Detection. *Database (Oxford)* **2015;***2015*:bav084. doi:10.1093/database/bav084.

30. Martin, W.; Rujan, T.; Richly, E.; Hansen, A.; Cornelsen, S.; Lins, T.; Leister, D.; Stoebe, B.; Hasegawa, M.; Penny, D. Evolutionary Analysis of *Arabidopsis, Cyanobacterial, and Chloroplast* Genomes Reveals Plastid Phylogeny and Thousands of Cyanobacterial Genes in the Nucleus. *Proc. Natl. Acad. Sci. USA* **2002,** *99*(19), 12246–12251.

31. Richly, E.; Leister, D. An Improved Prediction of Chloroplast Proteins Reveals Diversities and Commonalities in the Chloroplast Proteomes of *Arabidopsis* and Rice. *Gene* **2004,** *329,* 11–6.

32. Krishnan, N. M.; Rao, B. J. A Comparative Approach to Elucidate Chloroplast Genome Replication. *BMC Genomics* **2009,** *10*(237), 237.

33. Thomas, B. R.; Van Deynze, A.; Bradford, K. J. *Production of Therapeutic proteins in plants.* In Agricultural Biotechnology in California Series; Publication 8078, **2002.**

34. Gelvin, S. B. *Agrobacterium*-mediated Plant Transformation: The Biology Behind the "Gene-Jockeying" Tool. *Microbiol. Mol. Biol. R.* **2003,** *67*(1), 16–37.

35. Kondrák, M.; van der Meer, I. M.; Bánfalvi, Z. Generation of Marker- and Backbone-Free Transgenic Potatoes by Site Specific Recombination and a Bi-functional Marker Gene in a Non-regular One-border *Agrobacterium* Transformation Vector. *Transgenic Res.* **2007,** *15,* 729–737.

36. Bukovinszki, A.; Diveki, Z.; Csanyi, M.; Palkovics, L.; Balazs, E. Engineering Resistance to PVY in Different Potato Cultivars in a Marker-free Transformation System Using a 'Shooter Mutant' *A tumefaciens. Plant Cell Rep.* **2007,** *26,* 459–465.

37. Djukanovic, V.; Lenderts, B.; Bidney, D.; Lyznik, L. A. Site-specific Recombination in Maize Vesna Djukanovic et al. A Cre:FLP Fusion Protein Recombines FRT or loxP Sites in Transgenic Maize Plants. *Plant Biotech. J.* **2008,** *6,* 770–781.

38. Akella, V.; Lurquin, P. F. Expression in Cowpea Seedlings of Chimeric Transgenes After Electroporation into Seed-derived Embryos. *Plant Cell Rep.* **1993,** *12*(2), 110–117.

39. Hefferon, K. L. Nutritionally Enhanced Food Crops; Progress and Perspectives. *Int. J. Mol. Sci.* **2015,** *16*(2), 3895–3914.

40. Tang, M.; He, X.; Luo, Y.; Ma, L.; Tang, X.; Huang, K. Nutritional Assessment of Transgenic Lysine-rich Maize Compared with Conventional Quality Protein Maize. *J. Sci. Food Agric.* **2013,** *93*(5), 1049–1054.

41. Buhari, M. L.; Babura, R. S.; Vyas, N. L.; Badaru, S.; Umar, Y. H. Role of Biotechnology in Phytoremediation. *J. Biorem. Biodegrad.* **2016**, *7*, 2.

42. Richter, L. J.; Thanavala, Y.; Arntzen, C. J.; Mason, H. S. Production of Hepatitis B Surface Antigen in Transgenic Plants for Oral Immunization. *Nat. Biotechnol.* **2000**, *18*(11), 1167–1171.

43. Hood, E. E.; Woodard, S. L.; Horn, M. E. Monoclonal Antibody Manufacturing in Transgenic Plants—Myths and Realities. *Curr. Opin. Biotechnol.* **2002**, *13*, 630–635.

44. Yao, J.; Weng, Y.; Dickey, A.; Wang, K. Y. Plants as Factories for Human Pharmaceuticals: Applications and Challenges. *Int. J. Mol. Sci.* **2015**, *16*, 28549–28565. DOI:10.3390/ijms161226122.

45. Gong, Y.; Hu, H.; Gao, Y.; Xu, X.; Gao, H. Microalgae as Platforms for Production of Recombinant Proteins and Valuable Compounds: Progress and Prospects. *J. Ind. Microbiol. Biotechnol.* **2011**, *38*, 1879–1890.

46. Petruccelli, S.; Otegui, M. S.; Lareu, F.; Tran Dinh, O.; Fitchette, A. C.; Circosta, A.; Rumbo, M.; Bardor, M.; Carcamo, R.; Gomord, V.; et al. A KDEL-tagged Monoclonal Antibody Is Efficiently Retained in the Endoplasmic Reticulum in Leaves, but Is Both Partially Secreted and Sorted to Protein Storage Vacuoles in Seeds. *Plant Biotechnol. J.* **2006**, *4*, 511–527.

47. Svab, Z.; Hajdukiewicz, P.; Maliga, P. Stable Transformation of Plastids in Higher Plants. *Proc. Natl. Acad. Sci. USA* **1990**, *87*, 8526–8530.

48. Huang, F. C.; Klaus, S. M. J.; Herz, S.; Zuo, Z.; Koop, H. U.; Golds, T. J. Efficient Plastid Transformation in Tobacco Using the *AphA-6* Gene and Kanamycin Selection. *Mol. Genet. Genomics* **2002**, *268*, 19–27.

49. Maliga, P. *Towards Plastid Transformation in Flowering Plants*; TIBTECH Vol. 11; Elsevier Science Publishers Ltd.: **1993**, 101–107.

50. Stern, D. B.; Higgs, D. C.; Yang, J. Transcription and Translation in Chloroplasts. *Trends Plant Sci.* **1997**, *2*, 308–315.

51. Suzuki, H.; Ingersoll, J.; Stern, D. B.; Kindle, K. L. Generation and Maintenance of Tandemly Repeated Extrachromosomal Plastid DNA in *Chlamydomonas* Chloroplast. *Plant J.* **1997**, *11*(4), 635–658.

52. Oey, M.; Lohse, M.; Scharff, L. B.; Kreikemeyer, B.; Bock, R. Plastid Production of Protein Antibiotics Against Pneumonia via a New Strategy for High-level Expression of Antimicrobial Proteins. *Proc. Natl. Acad. Sci. USA.* **2009**, *106*, 6579–6584.

53. Seon, J.-H.; Szarka, J. S.; Moloney, M. M. A Unique Strategy for Recovering Recombinant Proteins from Molecular Farming: Affinity Capture on Engineered Oil Bodies. *J. Plant Biotechnol.* **2002**, *4*(3), 95–101.

54. Wandelt, C. I.; Khan, M. R.; Craig, S.; Schroeder, H. E.; Spencer, D.; Higgins, T. J. Vicilin with Carboxy-terminal KDEL Is Retained in the Endoplasmic Reticulum and Accumulates to High Levels in the Leaves of Transgenic Plants. *Plant J.* **1992**, *2*, 181.

55. Ma, JK-C.; Drake, P. M. W.; Christou, P. The Production of Recombinant Pharmaceutical Proteins in Plants. *Genetics* **2003**, *4*, 794–805.

56. Kamenarova, K.; Abumhadi, N.; Gecheff, K.; Atanassov, A. Molecular Farming in Plants: An Approach of Agricultural Biotechnology. *J. Cell Mol. Biol.* **2005**, *4*, 77–86.

57. McCormick, A. A. Tobacco Derived Cancer Vaccines for Non-Hodgkin's Lymphoma: Perspectives and Progress. *Hum Vaccin* **2011**, *7*(3), 305–312.

58. Ben Rejeb, I.; Pastor, V.; Mauch-Mani, B. Plant Responses to Simultaneous Biotic and Abiotic Stress: Molecular Mechanisms. *Plants* **2014**, *3*(4), 458–475.

59. Fraire-Velázquez, S.; Rodríguez-Guerra, R.; Sánchez-Calderón, L. *Abiotic and Biotic Stress Response Crosstalk in Plants-physiological, Biochemical and Genetic Perspectives*; Shanker, A., Ed.; InTech Open Access Company: Rijeka, Croatia, 2011; pp 1–26.

60. Laloi, C.; Appel, K.; Danon, A. Reactive Oxygen Signalling: The Latest News. *Curr. Opin. Plant Biol.* **2004**, *7*, 323–328.

61. Spoel, S. H.; Dong, X. Making Sense of Hormone Crosstalk During Plant Immune Response. *Cell Host Microbe.* **2008**, *3*, 348–351.

62. Zhu, J. K. Salt and Drought Stress Signal Transduction in Plants. *Annu. Rev. Plant Biol.* **2002**, *53*, 247–273.

63. Munns, R.; Tester, M. Mechanisms of Salinity Tolerance. *Annu. Rev. Plant Biol.* **2008**, *59*, 651 681.

64. Chinnusamy, V.; Jagendorf, A.; Zhu, J. K. Understanding and Improving Salt Tolerance in Plants. *Crop Sci.* **2005**, *45*, 437–448.

65. Karimi, G.; Ghorbanli, M.; Heidari, H.; Nejad, R. K.; Assareh, M. The Effects of NaCl on Growth, Water Relations, Osmolytes and Ion Content in Kochia Prostrata. *Biol. Plant.* **2005**, *49*, 301–304.

66. Burssens, S.; Himanen, K.; Van De Cotte, B.; Beeckman, T.; Van Montagu, M.; Inzé, D.; Verbruggen, N. Expression of Cell Cycle Regulatory Genes and Morphological Alterations in Response to Salt Stress in *Arabidopsis thaliana. Planta.* **2000**, *211*, 632–640.

67. Bray, E. A.; Shih, T. Y.; Moses, M. S.; Cohen, A.; Imai, R.; Plant, A. L. Water-deficit Induction of a Tomato H1 Histone Requires Abscisic Acid. *Plant Growth Regul.* **1999**, 29, 35–46.

68. Rensink, W. A., Iobst, S., Hart, A., Stegalkina, S., Liu, J., Buell, C. R. Gene Expression Profiling of Potato Responses to Cold, Heat, and Salt Stress. *Funct. Integr. Genomics* **2005**, *5*(4), 201–207.

69. Zhang, N.; Si, H.-J.; Wen, G.; Du, H.-H.; Liu, B.-L.; Wang, D. Enhanced Drought and Salinity Tolerance in Transgenic Potato Plants with a BADH Gene from Spinach. *Plant Biotechnol. Rep.* **2011**, *5*, 71–77.

70. Foyer, C.; Noctor, G. Redox Homeostasis and Antioxidant Signaling: A Metabolic Interface Between Stress Perception and Physiological Responses. *Plant Cell.* **2005**, *17*, 1866–1875.

71. Spoel, S. H.; Loake, G. J. Redox-based Protein Modifications: The Missing Link in Plant Immune Signalling. *Curr. Opin. Plant Biol.* **2011**, *14*, 358–364.

72. Kissoudis, C.; van de Wiel, C.; Visser, R. G. F.; van der Linden, G. Enhancing Crop Resilience to Combined Abiotic and Biotic Stress Through the Dissection of Physiological and Molecular Crosstalk. *Front. Plant Sci.* **2014**, *5*, e207.

73. Pastor, V.; Luna, E.; Ton, J.; Cerezo, M.; García-Agustín, P.; Flors, V. Fine Tuning of Reactive Oxygen Species Homeostasis Regulates Primed Immune Responses in *Arabidopsis. Mol. Plant Microbe Interact.* **2013**, *11*, 1334–1344.

74. Bowler, C.; Fluhr, R. The Role of Calcium and Activated Oxygens as Signals for Controlling Cross-tolerance. *Trends Plant Sci.* **2000**, *5*, 241–246.

75. Xia, X.-J.; Wang, Y.-J.; Zhou, Y.-H.; Tao, Y.; Mao, W.-H.; Shi, K.; Asami, T.; Chen, Z.; Yu, J.-Q. Reactive Oxygen Species Are Involved in Brassinosteroid-induced Stress Tolerance in Cucumber. *Plant Physiol.* **2012,** *158,* 1034–1045.

76. Miller, G.; Suzuki, N.; Ciftci-Yilmaz, S.; Mittler, R. Reactive Oxygen Species Homeostasis and Signaling During Drought and Salinity Stresses. *Plant Cell Environ.* **2010,** *33,* 453–467.

77. Brader, G.; Djamei, A.; Teige, M.; Palva, E. T.; Hirt, H. The MAP Kinase Kinase MKK2 Affects Disease Resistance in *Arabidopsis. Mol. Plant Microbe Interact.* **2007,** *20,* 589–596.

78. Li, Y.; Zhang, L.; Wang, X.; Hao, L.; Chu, X.; Guo, X. Cotton GhMPK6a Negatively Regulates Osmotis Tolerance and Bacterial Infection in Transgenic *Nicotiana benthaminana* and Plays a Pivotal Role in Development. *FEBS J.* **2013,** *280,* 5128–5144.

79. Cramer, G. R.; Urano, K.; Delrot, S.; Pezzotti, M.; Shinozaki, K. Effects of Abiotic Stress on Plants: A Systems Biology Perspective. *BMC Plant Biol.* **2011,** *11,* 163–177.

80. Ton, J.; Flors, V.; Mauch-Mani, B. The Multifaceted Role of ABA in Disease Resistance. *Trends Plant Sci.* **2009,** *14,* 310–317.

81. Todaka, D.; Nakashima, K.; Shinozaki, K.; Yamaguchi-Shinozaki, K. Toward Understanding Transcriptional Regulatory Networks in Abiotic Stress Responses and Tolerance in Rice. *Rice J.* **2012,** *5,* 1–9.

82. Foyer, C. H.; Karpinska, B.; Krupinska, K. The Functions of WHIRLY1 and REDOX-RESPONSIVE TRANSCRIPTION FACTOR 1 in Crosstolerance Responses in Plants: A Hypothesis. *Philos. Trans. R. Soc. Lond. B Biol. Sci.* **2014,** *369,* 20130226.

83. Rejeb, I. B.; Pastor, V.; Mauch-Mani, B. Plant Responses to Simultaneous Biotic and Abiotic Stress: Molecular Mechanisms. *Plants* **2014,** *3,* 458–475.

84. Seo, P. J.; Park, C.-M. MYB96-mediated abscisic Acid Signals Induce Pathogen Resistance Response by Promoting Salicylic Acid Biosynthesis in *Arabidopsis. New Phytol.* **2010,** *186,* 471–483.

85. Seo, P. J.; Kim, M. J.; Park, J. Y.; Kim, S. Y.; Jeon, J.; Lee, Y. H.; Kim, J.; Park, C. M. Cold Activation of A Plasma Membrane-tethered NAC Transcription Factor Induces a Pathogen Resistance Response in *Arabidopsis. Plant J.* **2010,** *61,* 661–671

86. Joshi, R.; Wani, S. H.; Singh, B.; Bohra, A.; Dar, Z. A.; Lone, A. A.; … Singla-Pareek, S. L. Transcription Factors and Plants Response to Drought Stress: Current Understanding and Future Directions. *Front. Plant Sci.* **2016,** *7,* 1029.

87. Younis, A.; Siddique, M. I.; Kim, C-K.; Lim, K-B. RNA Interference (RNAi) Induced Gene Silencing: A Promising Approach of Hi-tech Plant Breeding. *Int. J. Biol. Sci.* **2014,** *10*(10), 1150–1158.

88. Baulcombe, D. RNA Silencing in Plants. *Nature* **2004,** *431*(7006), 356–363.

89. Dinesh-Kumar, S. P.; Anandalakshmi, R.; Mrathe, R.; Schiff, M.; Liu, Y. Virus Induced Gene Silencing. *Methods Mol. Biol.* **2003,** *236,* 287–294.

90. Scorza, R.; Callahan, A.; Levy, L.; Damsteegt, V.; Webb, K.; Ravelonandro, M. Post-transcriptional Gene Silencing in Plum Pox Virus Resistant Transgenic European Plum Containing the Plum Pox Potyvirus Coat Protein Gene. *Transgenic Res.* **2001,** *10*(3), 201–209.

91. Boccara, M.; Sarazin, A.; Thiébeauld, O.; Jay, F.; Voinnet, O.; Navarro, L.; Colot, V. The *Arabidopsis* miR472-RDR6 Silencing Pathway Modulates PAMP- and

Effector-triggered Immunity Through the Post-transcriptional Control of Disease Resistance Genes. *PLoS Pathog.* **2014,** *10*(1), e1003883. http://doi.org/10.1371/journal.ppat.1003883.

92. Jiang, C. J.; Shimono, M.; Maeda, S.; Inoue, H.; Mori, M.; Hasegawa, M.; Sugano, S.; Takatsuji, H. Suppression of the Rice Fatty-acid Desaturase Gene OsSSI2 Enhances Resistance to Blast and Leaf Blight Diseases in Rice. *Mol. Plant Microbe Interact.* **2009,** *22*(7), 820–829.

93. Xin, M.; Wang, Y.; Yao, Y.; Xie, C.; Peng, H.; Ni, Z.; Sun, Q. Diverse Set of MicroRNAs Are Responsive to Powdery Mildew Infection and Heat Stress in Wheat (*Triticum aestivum* L.). *BMC Plant Biol.* **2010,** *10*, 123.

94. Mao, Y. B.; Cai, W. J.; Wang, J. W.; Hong, G. J.; Tao, X. Y.; Wang, L. J.; Huang, Y. P.; Chen, X. Y. Silencing a Cotton Bollworm P450 Monooxygenase Gene by Plant-mediated RNAi Impairs Larval Tolerance of Gossypol. *Nat Biotechnol.* **2007,** *25*(11), 1307–1313.

95. Baum, J. A.; Bogaert, T.; Clinton, W.; Heck, G. R.; Feldmann, P.; Ilagan, O.; Johnson, S.; Plaetinck, G.; Munyikwa, T.; Pleau, M.; Vaughn, T.; Roberts, J. Control of Coleopteran Insect Pests Through RNA Interference. *Nat Biotechnol.* **2007,** *25*(11), 1322–1326.

96. Yadav, B. C.; Veluthambi, K.; Subramaniam, K. Host-generated Double Stranded RNA Induces RNAi in Plant-parasitic Nematodes and Protects the Host from Infection. *Mol. Biochem. Parasitol.* **2006,** *148*(2), 219–222.

97. Yang, Y.; Jittayasothorn, Y.; Chronis, D.; Wang, X.; Cousins, P.; Zhong, G-Y. Molecular Characteristics and Efficacy of *16D10* siRNAs in Inhibiting Root-knot Nematode Infection in Transgenic Grape Hairy Roots. *PLoS One* **2013,** *8*(7), e69463. doi:10.1371/journal.pone.0069463.

98. Sunkar, R.; Zhu, J. K. Novel and Stress-regulated MicroRNAs and Other Small RNAs from *Arabidopsis. Plant Cell* **2004,** *16*(8), 2001–2019.

99. Jagtap, U. B.; Gurav, R. G.; Bapat, V. A. Role of RNA Interference in Plant Improvement. *Naturwissenschaften.* **2011,** *98*(6), 473–492.

100. Shinozaki, K.; Yamaguchi-Shinozaki, K. Molecular Responses to Dehydration and Low Temperature: Differences and Cross Talk Between Two Stress Signaling Pathways. *Curr. Opin. Plant Biol.* **2000,** *3*, 217–233.

101. Achard, P.; Herr, A.; Baulcombe, D. C.; Harberd, N. P. Modulation of Floral Development by Gibberellin-regulated MicroRNA. *Development* **2004,** *131*, 3357–3365.

102. Zhao, B. T.; Liang, R. Q.; Ge, L. F.; Li, W.; Xiao, H. S.; Lin, H. X.; Ruan, K. C.; Jin, Y. X. Identification of Drought-induced microRNAs in Rice. *Biochem. Biophys. Res. Commun.* **2007,** *354*, 585–590.

103. Jian, X.; Zhang, L.; Li, G.; Zhang, L.; Wang, X.; Cao, X.; Fang, X.; Zha, F. C. Identification of Novel Stress-regulated MicroRNAs from *Oryza sativa* L. *Genomics* **2010,** *95*, 47–50.

104. Liu, H. H.; Tian, X.; Li, Y. J.; Wu, C. A.; Zheng, C. C. Microarray-based Analysis of Stress-regulated MicroRNAs in *Arabidopsis thaliana. RNA* **2008,** *14*, 836–843.

105. Zhou, L.; Liu, Y.; Liu, Z.; Kong, D.; Duan, M.; Luo, L. Genome-wide Identification and Analysis of Drought-responsive MicroRNAs in *Oryza sativa. J. Exp. Bot.* **2010,** *61*, 4157–4168.

106. Lu, S. F.; Sun, Y. H.; Chiang, V. L. Stress-responsive MicroRNAs in *Populus. Plant J.* **2008,** *55*, 131–151.

107. Trindade, I.; Capitão, C.; Dalmay, T.; Fevereiro, M.; Santos, D. miR398 and miR408 Are Up-regulated in Response to Water Deficit in *Medicago truncatula*. *Planta* **2010,** *231*, 705–716.

108. Arenas-Huertero, C.; Pérez B, Rabanal, F.; et al. Conserved and Novel miRNAs in the Legume *Phaseolus vulgaris* in Response to Stress. *Plant Mol. Biol.* **2009,** *70*, 385–401.

109. Ding, D.; Zhang, L.; Wang, H.; Liu, Z.; Zhang, Z.; Zheng, Y. Differential Expression of miRNAs in Response to Salt Stress in Maize Roots. *Ann. Bot.* **2009,** *103*, 29–38.

110. Xin, M.; Wang, Y.; Yao, Y.; Xie, C.; Peng, H.; Ni, Z.; Sun, Q. Diverse Set of MicroRNAs Are Responsive to Powdery Mildew Infection and Heat Stress in Wheat (*Triticum aestivum* L.). *BMC Plant Biol.* **2010,** *10*, 123.

111. Jia, X.; Ren, L.; Chen, Q. J.; Li, R.; Tang, G. UV-B-responsive MicroRNAs in *Populus tremula*. *J. Plant Physiol.* **2009,** *166*, 2046–2057.

112. Lu, S. F.; Sun, Y. H.; Shi, R.; Clark, C.; Li, L. G.; Chiang, V. L. Novel and Mechanical Stress Responsive MicroRNAs in *Populus trichocarpa* That Are Absent from *Arabidopsis*. *Plant Cell* **2005,** *17*, 2186–2203.

113. Cañizares, M. C.; Nicholson, L.; Lomonossoff, G. P. Use of Viral Vectors for Vaccine Production in Plants. *Immunol. Cell Biol.* **2005,** *83*(3), 263–270.

CHAPTER 5

GENETIC MANIPULATION BY RECOMBINANT DNA TECHNOLOGY

CONTENTS

ABSTRACT

It has become possible to alter the genetic material and obtain the desired result through genetic manipulation techniques. Genetic engineering involves the manipulation of genes of an organism to express new desired traits. Bacteria, plants, yeast, and animals are common examples of genetically engineered organisms. Genetic manipulation involves physically moving and expressing a gene from a source organism to a recipient organism. In this chapter, we will briefly discuss the processes of cloning, and expressing the foreign gene through a plasmid in a host through heterologous gene expression to produce recombinant products. We will also briefly discuss the ethics behind the use and production of these products and genetically engineered organisms.

5.1   GENETIC ENGINEERING TECHNIQUES: THE BASIC TENET

Genetic manipulation techniques entail the alteration of genetic material to attain a predetermined result. Humans have been inadvertently manipulating genes ever since they began domesticating animals and raising crops, for example, mating of a bull with a cow from a high milk-yielding stock in the hope that the resulting offspring will inherit the high milk yield characteristics. This has been done for thousands of years without any knowledge of genes or the mechanism of inheritance. The other most obvious example is wheat, whose genes were manipulated by cross-breeding. Crossing of a wheat variety with wild grass gives hybrid wheat. Hybrid wheat was crossed with wild grass to give hybrid wheat (see Chapter 4, Fig. 4.1) used for making flour and bread. This method of genetic recombination can take place only between varieties of the same or closely related species.

Genetic engineering makes it possible to transfer genes. The genes can also be transferred from one species to a totally different species. There are several ways in which genes from one organism can be inserted into a different organism.

Paul Berg, in 1972, combined DNA from the cancer-causing monkey virus SV40 with that of the virus lambda[1] to create the first recombinant DNA molecules, but terminated the experiment before it could be taken any further when he realized the dangers of his experiment. He later worked on the safety concerns of the recombinant DNA research. Berg later continued his recombinant DNA research and was awarded the 1980 Nobel Prize in chemistry.[1]

The genetic engineering technique was further developed by Cohen and Boyer. In 1973, Cohen researched in the field of bacterial plasmids.[2,3] He studied how the genes of plasmids could make bacteria resistant to antibiotics and combined with the results obtained by Herbert Boyer, led to the development of methods to combine and transplant genes. They observed that the genome of a particular cell can be altered by this recombinant DNA technology to alter its phenotype in the desired fashion. In their preliminary genetic engineering experiments, in 1973, they showed that the gene for frog ribosomal RNA could be transferred and expressed in a bacterial cell. They first constructed a plasmid pSC101, which they called the vector. This plasmid contained a gene for tetracycline resistance and a single site for the restriction enzyme *EcoRI*. The frog DNA was cleaved into small segments by *EcoRI*. The frog DNA fragments were then combined (or ligated) with the plasmid, which had also been cleaved with *EcoRI*. A strain of *Escherichia coli* was then transformed with the ligated plasmids and plated onto a growth medium containing tetracycline. The cells that contained the plasmid carrying and expressing the gene for tetracycline resistance grew and formed colonies on that medium. The colonies that formed after growth were tested for the presence of frog ribosomal RNA gene and some of them were actually positive.

5.1.1 ISOLATION OF DNA

Isolation of the gene coding for the required protein: Genes are specific sequences of DNA that contain information to direct the synthesis of proteins. Proteins perform a range of functions from enzymes to structural components of a cell.[4]

5.1.2 USING RESTRICTION ENZYMES

To fragment the DNA and screen for the gene of interest, restriction enzymes are required. Genes can be cut at specific sites, at sequences recognized by proteins also known as restriction enzymes. Each enzyme recognizes and cuts DNA at a particular stretch of sequence. The restriction enzymes are of two major types: *type* I and *type* II.[5,6] The type I enzymes recognize specific sequence, but cut elsewhere and type II enzyme cuts only within the recognition site. They are extremely accurate; they only cut at their specific recognition sequence. This confounding property allows us to cut DNA into fragments that can be isolated, separated, and/ or analyzed. A restriction enzyme cuts double-stranded DNA at a specific recognition nucleotide sequences (A, T, C, and G) known as restriction sites. These enzymes are found in bacteria and are thought to have evolved as a defense mechanism against invading viruses. Restriction enzymes are required for removing or replacing genes. A very commonly used restriction enzyme in recombinant DNA technology is *EcoR*I, isolated from *E. coli* (Fig. 5.1), which generates a "sticky end," generating 5′ overhang as it cuts the DNA.[7] There are a few restriction enzymes that generate "blunt ends" as they act, for example, *Alu*I enzyme. In addition, since these enzymes recognize specific sequence for action, the number of cuts made in the DNA of an organism is limited and would generate a more or less repetitive pattern.

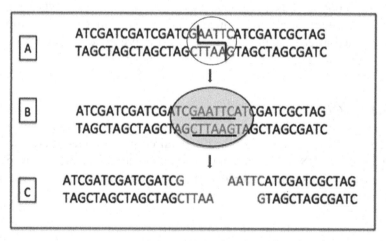

FIGURE 5.1 *EcoR*I is a very commonly used restriction enzyme in recombinant DNA technology. It is isolated from *E. coli*, and generates a "sticky end." (A) Palindromic, complementary recognition sequence of restriction enzyme *EcoR*I (in circle), with its cut pattern. (B) *EcoR*I binds to the recognition sequence. (C) The enzyme action results in linear DNA with five nucleotide sticky ends with 5′ end overhangs of AATTC.

5.1.3 CLONING

In cloning, the desired gene is inserted into a suitable vector (plasmid/ cosmid/phage/transposons/yeast artificial chromosome (YAC)/bacterial artificial chromosome (BAC), etc.) forms recombinant DNA and is subsequently introduced into a suitable competent "host" organism.

5.1.3.1 DNA LIGATION

Two nucleic acid fragments are joined by using enzyme ligase.[8] Phosphodiester bonds is formed between the 3'-hydroxyl of one DNA terminus and the 5'-phosphoryl of another to join the ends of DNA fragments together. ATP or NAD$^+$ usually act as cofactors in the reaction. The factors affecting ligation are DNA concentration, temperature, ligase concentration, and buffer composition. T4 DNA ligase is the most commonly used ligase.[9]

For a *"sticky-end"* (restriction enzymes generate a 4-base single-stranded overhang called the sticky or cohesive end) ligation, the ends can anneal to complementary, compatible ends. It is advisable to use two different restriction enzymes to digest the DNA so that different ends are generated for the insertion of a DNA fragment into a vector. The advantages include generating DNA fragments with two different ends can prevent the religation of the vector without any insert, and it would also allow a directional insertion of the fragment of interest.

For *"blunt-end"* ligation, any blunt end may be ligated to another blunt end because it does not involve base-pairing of the overhangs. Moreover, blunt-end ligation is less efficient than the sticky-end ligation. Usually, the vector needs to be dephosphorylated to minimize self-ligation if both ends of the fragment are blunt-ended.

If two different restriction sites cannot be used, the vector DNA may need to be dephosphorylated to avoid a high background of vector DNA that has re-circularized itself without the insert. Calf-intestinal alkaline phosphatase (CIAP) removes the phosphate group from the 5' end of digested DNA and is commonly used for dephosphorylation.[10]

5.1.3.2   DNA RECOMBINANT

The DNA fragments cannot function on their own and must be part of the genetic material of living things cells before the genes they contain can be activated. Thus, the next step of genetic engineering involves putting the DNA fragment into a host cell. DNA fragments that have been cut by restriction enzymes may be combined with "vector DNA" so that they can later be inserted into a host cell. Bacteria also may contain DNA molecules called *plasmids* that are small, circular, and can be removed and cut with a restriction enzyme. The cut ends are ligated to the foreign DNA fragment, and the formation of a recombinant DNA molecule is allowed.[11] The new DNA strand created would not normally occur in nature and is therefore considered to be artificial. The nonhuman organisms have thus been used to make a recombinant biological molecule such as human insulin.

5.1.3.3   DNA INSERTION, SELECTION, AND AMPLIFICATION

During the two initial steps of genetic engineering, DNA fragments containing the desired gene are obtained and inserted into DNA vector that has to be inserted into the recipient cell (plasmid, in the case of bacteria), forming recombinant DNA. It is easiest to use bacteria amplifying the DNA. Bacteria are made "competent" to eventually take up the DNA in its own DNA.[12] This is considered as DNA insertion. These new bacteria are then cultured in a selection medium. This process is called as DNA cloning. Cloning has also been successful in other organisms such as plants and animals.[13] It is now possible to insert genes from one organism into another. Such organisms are also known as "transgenic," meaning containing the gene from across species. Generating transgenics requires many of the already mentioned genetic engineering techniques and allows us to create organisms with foreign traits.

5.1.3.4   SCREENING OF THE HOST ORGANISMS

Antibiotic selection marker: Vectors contain genes that confer resistance to them when the transformed organism grows on the medium containing that antibiotic. This ensures successful insertion of a desired DNA fragment in the vector.

Insertional inactivation: Insertional inactivation is another common procedure for the detection of successful insertion.[14] For example, if an insert is present in between the genes for antibiotic resistance, the gene gets inactivated and does not confer resistance to the organism.

Gel electrophoresis: The DNA isolated from the recombinant clone is cut with restriction enzymes into fragments to confirm if the recombinant DNA has the same DNA insert and the vector with which the cloning experiment started. Different charges are placed at either end of an electrophoresis gel containing tray. DNA has a slight negative charge, so when the DNA is placed into the tray, it will slowly move across the gel (toward the positive charge) and the fragments move at different speeds because they are of different sizes, that is, the larger the fragment, the slower it will move, as shown in Figure 5.2.

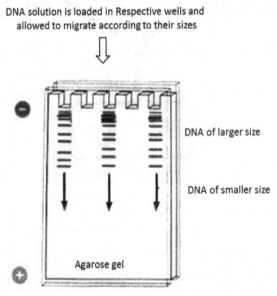

FIGURE 5.2   Agarose gel electrophoresis. DNA fragments are loaded along with a dye (to visualize their movement) into the wells. The DNA being negatively charged is attracted to the positive electrode. The shorter DNA fragments move faster than the longer fragments.

Sequencing: While sequencing the DNA fragment, the DNA nitrogenous bases (ATCG) are identified and read along the length of the DNA fragment. In sequencing reactions, only one strand of the double helix is used to sequence the DNA. The nitrogenous bases are tagged using different chemical treatments that break the DNA fragment into smaller pieces and reveals the base sequence. The tagging of the bases with different moieties helps to detect the order of A, T, C, and G, enabling to read the sequence in which the bases are arranged.

Once a clone of cells having the desired gene is confirmed, unlimited amounts of desired DNA can be prepared for further manipulations. The schematic of cloning process has been described in Figure 5.3.[15]

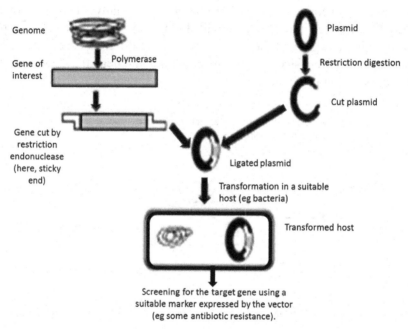

FIGURE 5.3   Flow diagram showing cloning schematics of cloning in bacteria.

5.2   CLONING VECTORS

These are certain DNA molecules that can transfer the desired genetic material into another cell. (The principle is similar to transmission of pathogens to humans via mosquito vector).[16,17]

Insertion of the exogenous DNA (gene of interest and vector) into prokaryotic cells is called transformation, while in eukaryotes, the process is called transfection. The usage of a viral vector is sometimes called transduction.

Vectors require to possess following properties:

1. Ori—a replication origin that ensures that the vector can replicate and be amplified in the host (see Chapter 3, "Replication and
   Maintenance of Bacterial Plasmid" under Section 3.1.2.1).
2. Restriction endonuclease recognition and cleavage sites to enable
   the insertion of the desired gene.
3. Possess "selection markers" that would express in the host cells
   and permit identification of the host cells that have taken up the
   desired gene, that is, successfully transformed cells. A common
   marker is an antibiotic selection gene that imparts antibiotic resistance to the host cells that take up the vector with the gene of
   interest against the nontransformed cells that remain susceptible to
   the antibiotic.
4. The introduction of the vector in the host should be easy and vector
   isolation should be easy.

5.2.1   TYPES OF VECTORS

5.2.1.1   PLASMIDS

Plasmids are extrachromosomal, circular, double-stranded DNA molecules
that are capable of autonomous replication. They are present in bacteria,
yeast, and sometimes in plant and animal cells. The term "plasmid" was
coined by Joshua Lederberg in 1952. Bacterial plasmids usually impart
certain useful metabolic characters to the host organism to overcome
unfavorable conditions such as antibiotic resistance. The cells that have a
plasmid would survive the antibiotic while the cells that lack the plasmid
would be antibiotic sensitive (Fig. 5.4).

A typical plasmid has:

1. An origin of replication.
2. A multiple cloning site (MCS)—a cluster of sites that are recognized and cleaved by restriction enzymes. This site permits the
   plasmid to be cut with specific enzymes used for isolating the gene
   of interest that later enables the joining (ligation) of the two fragments namely, the vector and the gene of interest.

3. A marker gene—this could be antibiotic resistance. For example, a
    plasmid with tetracycline resistance would impart the transformed
    cells to be resistant to tetracycline. The nontransformed cells that
    do not take up exogenous DNA would be sensitive to tetracycline
    and would not grow.

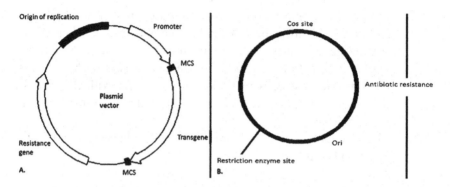

FIGURE 5.4   Diagram of plasmid and cosmid vectors.

Another common marker is the β-galactosidase (an enzyme that enables
growth on lactose medium). If the vector ligates with the desired gene, the
β-galactosidase gene gets interrupted and hence transformed host cells are
unable to utilize lactose-producing white colonies on an X-gal (a chro-
mogenic substrate) plate, whereas nontransformed colonies hydrolyse the
X-gal to produce blue colonies.[18]

Plasmids are named with a lowercase p (p) followed by letters in the
researcher(s)' names and having numbers given by the scientists(s). For
example, pUC19 is a plasmid discovered from University of California.[19]

Plasmids are easy to use but cannot accommodate large DNA fragments (greater than 10 kbp length). Hence, the shift toward the use of
cosmids as cloning vectors.

5.2.1.2   COSMIDS

These are used for cloning of large DNA molecules. Cosmids contain the
elements of plasmids along with "cos sites" from λ phages.[20] These cos

sites are used by the phage for circularizing and packing the DNA in the host cells. These cosmids can accommodate up to 45 kbp of DNA. The vector does not contain λ phage proteins hence, the transformed cells do not die. The gene of interest is inserted into the cosmid using restriction enzyme sites. The exogenous DNA then circularizes in the host cells that can be selected using markers similar to that of the plasmids. These vectors can accommodate DNA fragments of sizes ranging from 35 to 50 kb.

5.2.1.3   TRANSPOSONS

These are sequences of DNA that can transpose or shift from one place in the genome to another. The term was used by Hedges and Jacob in 1974.[21] In bacteria, transposons are known to shuttle between the genome and plasmids allowing for addition of marker genes to the genome. In primates, we can see *Alu* sequences that are 300 bases long and account for repetitive DNA sequences. Alu elements are the most abundant transposable elements, containing more than million copies scattered throughout the human genome.[22] Alu elements are classified as short interspersed nuclear elements (SINEs) among the class of repetitive DNA elements when they are nearly 300 base pairs long.

Due to their property of shifting their location from one place to another in the genome, they can produce gene activation/inactivation or deletion or illegitimate recombination in the genome. Hence, they are called jumping genes and have played a role in evolution. The organization and recombination in transposition has been depicted in Figure 5.5. and transposition using transposase is shown in Figure 5.6.

5.2.1.3.1   Types of Transposons

Sleeping beauty (SB) is the *Tc1/mariner family* of transposons that is widely present in many taxa such as rotifers, fungi, fish, and mammals.[23] SB is a promising system for gene transfer in vertebrates, stem cells, and cell lines. It has been used in human T cells with a stable transfer. It has also been used in screens for identifying cancer and insertional mutagenesis. Other transposons from this family include Frog prince—from the northern leopard frog. It has been used efficiently in gene trapping.

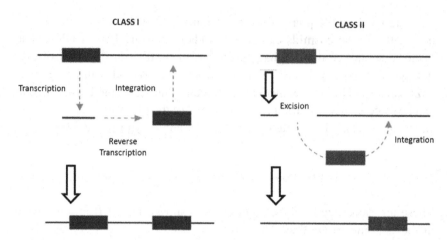

FIGURE 5.5 Organization of repeated DNA sequences and the mechanism of recombination in transposition. Class I—transcribe their DNA into RNA, and are thus called retrotransposons. The RNA is converted back to a new copy of DNA by reverse transcription and inserts itself in another location in the genome. Class II—These transposons use the cut-and-paste mechanism, in which the enzyme transposase make a cut at a target site, cuts out the transposable element, and ligates it to the new position.

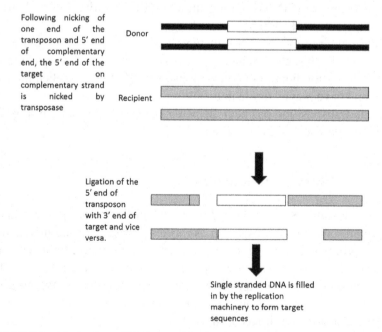

FIGURE 5.6 Mechanism of DNA recombination in transposition using transposase.

DNA transposons from *the family Hat* (hobo/Ac/Tam3) from eukary-otes have been used as genetic tools.[23] For example, Tol2 has been used in Xenopus, cultured human cells including stem cells.

Transposons are inexpensive but are less efficient for gene transfer when compared to viruses. However, the new transposon vector called *piggyBac* superfamily has been found to be at par with viral vectors. It is a DNA transposon identified in the genome of the Cabbage Looper moth (*Trichoplusiani*). Its transposition mechanism, control, and life cycle is similar to that of Tc1/*mariner* elements.[23]

5.2.1.4 BACTERIOPHAGES

These are viruses that infect bacteria. Bacteriophages show two types of life cycles: lytic, where new virus particles produced are released from the host bacteria causing host death and lysogenic, where the virus integrates itself in the host genome (see Chapter 3, Figures 3.6 and 3.10). Large DNA molecules can be cloned by phages. Generally, phages contain a protein coat externally, enclosing DNA/RNA as genetic material. M13 (Fig. 5.7) and λ phages (common host *E. coli*) (Fig. 3.11) are commonly used as tools (Chapter 3, "Bacteriophages," under Section 3.1.1).

From the genome of the phage, the portion encoding virulence is removed to enable cloning of the gene of interest. The genes for packaging the phage/gene into the protein coat are retained.

5.2.1.4.1 Phage λ as Vector

The phage has DNA of length 48.5 kbp. There are 12 bp "cos" sites at the end use for packaging the nucleic acid into protein coats. These cos sites cause the DNA to become circular in the host cells for packing (see Chapter 3, Section 3.11, 3.12).

As a genetic tool, up to 20 kbp DNA can be removed from the genome to accommodate a gene of interest. The recombinant DNA can be packaged within viral particles following which they are allowed to infect cultured bacterial host cells. A phage is a lytic organism that causes lysis of the host cells. Hence, we can see the recombinants as clear patches of no growth called "plaques" on the dense "lawn" of bacteria that are not recombinant.

5.2.1.4.2 Phage M13 as Vector

The phage M13 has a single-stranded DNA genome packed within a rod-shaped capsid (Chapter 3, "Bacteriophages," under Section 3.1.1; Fig. 5.7). These phages infect host F⁺ cells (cells that possess F pili). The DNA exists as both single- and double-stranded forms and hence are a source of single-stranded DNA. M 13 phages do not kill the host cells by lysis. Instead, the phages are produced without lysis and we can see slow growth of the host. Hence, the recombinant host cells are seen as turbid plaques on the bacterial lawn.

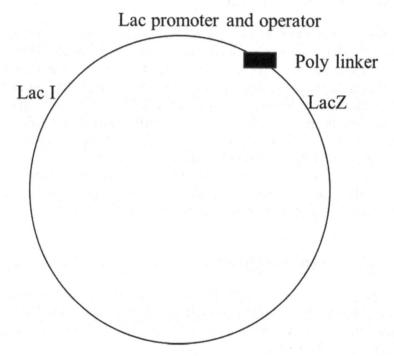

FIGURE 5.7 Diagram of an M13 phage vector. The replicating form of M13 is cleaved in the polylinker site that contains two restriction enzymes which can be used to clone the desired DNA fragment. DNA fragment is also cleaved with the same enzymes, ensuring cloning in the desired and single orientation.

These vectors have been modified as M13 mp18/19 that contain the lacZ site from the plasmids pUC 18/19. But, they cannot clone more than 2 kb of DNA. They have been modified as phagemid vectors, for example, pEMBL 18/19-0. These are the plasmids with an M13 origin of replication along with helper phages. Improved varieties such as Blue-Script (Stratgene) are available with added features such as different cloning strategies.

There are many requisites of cloning of single-strand DNA, such as for sequence analysis. The use of M13 vector makes it possible.

5.2.1.5 YEAST ARTIFICIAL CHROMOSOME

These are vectors for large DNA fragments in a yeast host. They were made by joining entities essential for the replication of yeast chromosomes with the DNA of interest.[24] They possess telomere sequences, centromere, and autonomous replicating sequences (ARS). The markers used here are those that help the transformed cells to grow on media that do not have certain nutrients. Sometimes, the YAC may get unstable while cloning large DNA (Fig. 5.8).

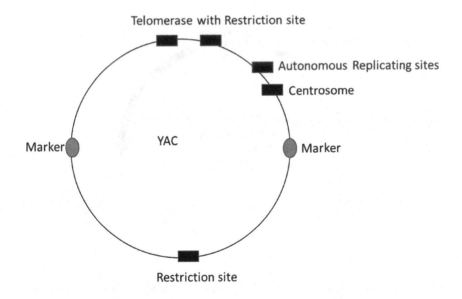

FIGURE 5.8 A YAC vector.

The sequences that function as centromeres (CEN) help in the attachment of microtubules during cell division such as mitosis and meiosis.

All vectors used in yeast can also replicate in *E. coli*. Hence, the vectors can first be cloned in *E. coli*, purified and then transform yeast. Hence, they are called shuttle vectors.

5.2.1.6   BACTERIAL ARTIFICIAL CHROMOSOME

BAC have been constructed from the F plasmid in *E. coli* used for transforming and cloning in bacteria and ensure that they are evenly distributed after bacterial cell division because they contain partition genes (Fig. 5.9). They can accommodate 350 kbp of DNA.[25] BACs often find use in genome projects to sequence the genome of organisms. BACs enable the amplification the short piece of the organism's DNA as an insert in, and then sequenced. The BACs constitute: a selectable marker (usually antibiotic resistance or *lacZ* gene for blue-white selection; T7 and SP6 phage promoters; *parA* and *parB* for partitioning of the F plasmid to daughter cells; *repE* for copy number maintenance and plasmid replication.

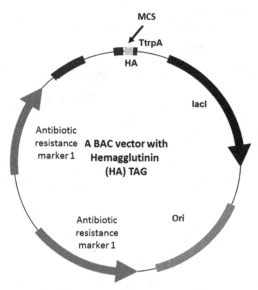

FIGURE 5.9   Diagram of a BAC vector with HA tag. This tag ensures easy purification of fusion proteins and allows detection of fusion proteins in immunoblotting, as well as, it is helpful for cell-localization studies. *eryr, ampr*: antibiotic resistance markers.

Artificial chromosomes have been reported in successfully transferring human antibody genes to cattle. The stability was satisfactory.[26] Additionally, they possess a selectable marker and a restriction site for inserting the gene of interest. They are easy to transform host cells, produce large amounts, and are easy to purify.

5.2.1.7   EUKARYOTIC EXPRESSION VECTORS

Eukaryotes show certain cellular mechanisms that entail the use of vectors that process and express the protein of interest.

First, eukaryotic proteins find difficulty in protein folding and expression due to different disulfide bonds patterns. Second, many cells splice (excise) out the introns (noncoding) from the exons (coding) regions of a gene. This is essential to produce functional proteins. Third, they need to possess promoters (to initiate the mRNA formation), polyadenylation sites, and transcription termination sites (Fig. 5.10).

They also need to possess origins of replication for propagation in *E. coli* and the eukaryotic cell along with a MCS (Fig. 5.10).

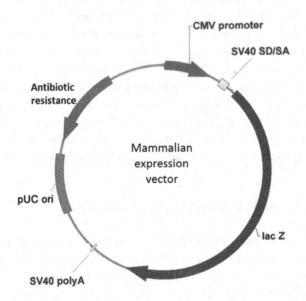

FIGURE 5.10   The diagram of a mammalian vector. The vector contains β-galactosidase enzyme expression (through lac Z) and SV40 late polyadenylylation signal. This vector also contains an antibiotic selection marker.

The vector can possess certain tags to ensure the successful translation of the coded mRNA. It has been found that incorporating the Kozak
sequence (ACCAUGG)[27] near the initiating codon AUG enhances the
success of protein expression. Apart from the stop/nonsense codons, signal
sequences can be introduced for the successful packing and secretion of
the protein. The mRNA stability is found to be enhanced with certain 5'
and 3' untranslated regions.

5.2.1.8   COMMONLY USED CLONING VECTORS

Besides cloning involving restriction enzymes, there are other technologies
that are very widely used in laboratories around the world. A few of them are:

5.2.1.8.1   Ligation Independent Cloning

This technique employs the 3'–5' exonuclease activity of T4 DNA polymerase in order to create 5' overhangs on both the vector and insert.[28] T4
polymerase will continue to function as an exonuclease in the presence of
any free, single deoxyribonucleotides (dNTP), until a base is exposed on
the single strand overhang which is complementary to the free nucleotide.
In the presence of exposed single strand base, T4 polymerase adds back the
free base, and is stalled at this point (because there are no other free bases
to add). For this purpose, complementary overhangs based on the choice
of restriction site and destination vector sequence are built into the polymerase chain reaction (PCR) primers for the insert. Ligation is not necessary
because long stretches of bases are already paired in the annealed product.
The nicks of the product are repaired by the replication process in the *E. coli*.

5.2.1.8.2   Gateway Recombination Cloning

This technique allows transfer of DNA fragments between different cloning
vectors, keeping intact the reading frame.[29] Using Gateway, DNA segments
can be cloned or subcloned for functional analysis. The initial step requires
insertion of a DNA fragment into a plasmid with two flanking recombination sequences called "att L 1" and "att L 2," to develop a "Gateway
Entry clone." Life science research scientists have widely adopted this

technology, especially for applications that require transfer of thousands of DNA fragments into one type of plasmid. "Gateway attB1, and attB2" sequences are added to the 5′, and 5′ end of a gene fragment, respectively, using gene-specific PCR primers and PCR-amplification.[30] This followed by mixing of PCR amplification products with special plasmids called "donor vectors" and an enzyme mix. Recombination and insertion of the PCR product containing att B sequence is catalyzed by the enzyme mix into the donor vector att P recombination sites. Once the cassette is part of the target plasmid, it is called a Gateway "Entry clone," and recombination sequences are referred to as the Gateway "att L" type. The gene cassette in the entry clone can then be simply and efficiently transferred using an enzyme mix into any "destination vector" (any Gateway plasmid that contains Gateway "att R" recombination sequences and elements such as promoters and epitope tags, but not ORFs). This technology is commercially available for use (see Chapter 3, Section 3.2.4; Fig. 3.13).

5.2.1.8.3 Isothermal Assembly Reaction

Isothermal cloning is based on the properties of three common enzymes.[31] The 5′ exonuclease, which digests the 5′ end of double-stranded DNA fragments to generate 3′ single-stranded overhangs in a reaction mixture that includes two or more DNA fragments with a homology of 20–40 bp at their ends. The resulting "sticky ends"—similar to that created by restriction enzymes, but with greater length of complementarity—find each other and anneal. Any remaining regions of single-stranded DNA are filled by the polymerase, and the ligase then fuses the nicks, resulting in a single DNA fragment. isothermal assembly reaction allows assembly of multiple DNA fragments in chosen orientation. Moreover, this method does not require any sequence (restriction enzyme sites or Gateway recombination sites) at the junctions.

5.2.1.8.4 TA Cloning

The principle of TA cloning is based on the fact that *taq polymerase* leaves one adenosine (A) overhang on the 3′ end of PCR products, resulting in efficient hybridization between the 3′ A overhang of the PCR product and the 5′ T overhang of the vector backbone.[32] However, the 3′ A overhangs can degrade over time, further reducing ligation efficiency. TA cloning is generally used as the first step in the cloning of PCR products, before subcloning into another vector using restriction enzyme digestion whose sites are present on either side of the insert.

5.2.1.8.5   Type IIS Assembly

In this system, unique properties of *type IIS* restriction endonucleases are used, which cuts double-stranded DNA at a specified distance from the recognition sequence.[33] Thus, custom overhangs can be created in this system, which is not possible with traditional restriction enzyme cloning.

5.2.1.8.6   Oligonucleotide Stitching and Yeast-mediated Cloning

Yeast-mediated cloning takes advantage of the powerful recombination abilities of yeast, a technique very similar in principle to that of isothermal assembly reaction, except the in vitro reaction with purified enzymes in the latter.[34] In addition, two (or more) fragments of dsDNA with 30 or more bases of overlapping homology can be efficiently fused. Much larger final products compared to that in the other cloning methods can be generated (up to 100 kb), that utilize bacteria. Moreover, through oligonucleotide stitching, pieces of DNA that share no end homology can still be fused together. To accomplish this, two (or more) fragments of DNA that need to be fused, custom ordered DNA oligos of 60–80 bp bearing 30–40 bp of homology to the ends of the two (or more) fragments to be fused are introduced in yeast.[34]

5.2.1.8.7   Shuttle Vectors

These vectors can replicate in different host organisms. These vectors can be manipulated (mutagenesis, PCR, and cloning) and amplified in one organism (*E. coli*)[35] and then be used in a system in which the expression of the cloned gene is to be studied (e.g., yeast, as shown in Fig. 5.11).[36]

One of the most common types of shuttle vectors is the yeast shuttle vector. Almost all commonly used *S. cerevisiae* vectors are shuttle vectors. These vectors have two origins of replication (active in two different suitable hosts) and two selective markers (e.g., *trp* is used for detection in yeast and tet-r for detection in *E. coli*).

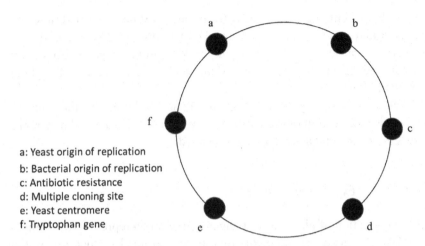

a: Yeast origin of replication
b: Bacterial origin of replication
c: Antibiotic resistance
d: Multiple cloning site
e: Yeast centromere
f: Tryptophan gene

FIGURE 5.11 A yeast shuttle vector.

5.2.1.9   DNA LIBRARIES

These are a collection of genes from a particular organism that is stored and maintained in a microbial population. They are of two types:

5.2.1.9.1   Genomic Library

It represents all the clones of DNA sequences that represent the complete genome of an organism. Here, the entire genome (DNA) of an organism is isolated by cell lysis, ultracentrifugation, and electrophoresis or chromatography and digested into fragments using restriction enzymes. These restriction fragments are inserted into suitable vectors and transferred into a suitable host such as bacteria/yeast.

Whenever a desired gene is to be screened, complementary radioactive probes can be used.

5.2.1.9.2   cDNA Library

cDNA means complementary DNA. The genes of eukaryotes consist of coding (exons) and noncoding (introns) regions. Only the exons are transcribed into mRNA and translated subsequently into proteins.

In this technique, mRNA of cells at various time intervals/frames are isolated and reverse transcribed into DNA (cDNA). This entails the use of reverse transcriptase enzyme. The cDNA can be prepared from every protein (hence, mRNA) and inserted into a suitable vector and cloned into a suitable host.

The cDNA library differs from the genomic library as the genomic library is a collection of the entire genome while the cDNA library represents the genes that are actively transcribed.

5.2.1.9.3 Expression Library

It is a collection of plasmids or phages containing a representative sample of cDNA or genomic fragments that are constructed in such a way that they will be transcribed and translated by the host organism.[38] This library not only contains the DNA fragments of interest but can actually manufacture the protein coded by the fragment so that it may be detected by the antibody. For this, the cDNA fragment within the vector is inserted downstream of a bacterial promoter that causes the inserted fragment to be expressed. The protein expressed by the gene of interest is identified by protein activity or by western blotting using specific antibody. Figure 5.12 schematically compares the construction of genomic and cDNA libraries.

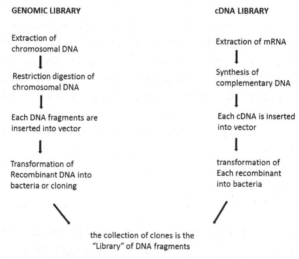

FIGURE 5.12   Construction of genomic library vs. cDNA library.

5.3 SCREENING FOR THE RIGHT RECOMBINANT

5.3.1 HYBRIDIZATION TECHNIQUES

5.3.1.1 SOUTHERN BLOTTING

Molecular biologist E. M. Southern developed this technique that allows for the screening and study of a cloned gene.[39] The DNA of choice is digested with restriction enzymes. The digested fragments can be separated on the basis of their size by agarose gel electrophoresis. These fractionated digest products are then transferred to a nitrocellulose membrane to develop an exact print or copy of the gel. Hence, the membrane is now a replica of the gel pattern. Now, a "probe" (suitable mRNA/DNA) is used to identify a particular gene/DNA sequence(s) of interest. Usually, the probe is labeled radioactively with P^{32}. This enables subsequent identification using an autoradiograph. The main use of the blot is identification of restriction patterns, information as to the number of copies of a gene in a genome and also comparative studies.

5.3.1.2 NORTHERN BLOTTING

It is a method for the detection of a specific mRNA in a mixture of RNAs. Similar to Southern blotting, the method is highly specific and sensitive. RNA from a cell/tissue is extracted and separated by electrophoresis. As in Southern blotting, the RNA is transferred to a nitrocellulose membrane and incubated with a labeled probe that is complementary to the RNA. Northern blotting is a standard for the direct study of gene expression at the level of mRNA, detection of mRNA transcript size, studies involving RNA degradation and RNA splicing, and often used to confirm and check to determine the half-life of RNA.

5.3.1.3 WESTERN BLOTTING

Also known as immunoblotting, western blotting is a well-established and widely used technique for the detection and analysis of proteins.[40] The method is based on binding between an antibody–protein complex via specific binding of antibodies to proteins immobilized on a membrane and detecting the bound antibody. A prerequisite for Western blotting the

SDS PAGE of the protein mixture. The Western blotting method was first described in 1979 by Towbin et al. and has since become one of the most commonly used methods in life science research. Western blotting is a very specific test and can detect one protein in a mixture of any number of proteins and also gives information about the size of the protein.

5.3.1.4 MICROARRAYS

All the mRNAs transcribed in an organism (transcriptome) can be completely analyzed by DNA microarray technique (see Chapter 4, Section 4.2.3.4 "Microarrays," Fig. 4.9). In this method, the isolated mRNA is converted to cDNA, labeled with a fluorescent dye and hybridized to a gene chip containing oligonucleotide sequences representing all or a subset of genes in that organism. The quantitation of fluorescence intensity of the cDNA bound to each probe gives the amount of mRNA expressed from each gene. DNA microarray data can be analyzed to identify clusters of genes with related functions that are similarly regulated under certain conditions (e.g., stress and disease) or under different growth conditions.

5.3.2 PCR OR GENE AMPLIFICATION

This is a technique used to generate or amplify manifold copies of a target or particular gene/DNA sequence of interest. It was invented by Kary Mullis.[41]

The reaction involves three basic steps in a thermal cycler that constitute one cycle of amplification. The cycles are repeated to obtain an exponential yield of the target DNA (Fig 5.13).

5.3.2.1 REQUIREMENTS FOR PCR

1. the target DNA segment (100–35,000 bp)
2. synthetic oligonucleotides called primers—forward and reverse that are needed to initiate the DNA copying and hence amplification
3. dNTP for constructing the new DNA copies—dATP, dTTP, dGTP and dCTP—these are the deoxy adenosine, thymine, guanosine, and cytosine triphosphates that are the backbone of DNA
4. thermostable DNA polymerase (Taq polymerase)

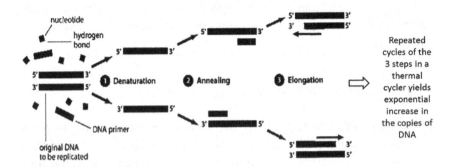

FIGURE 5.13 Schematics of PCR reaction.

The steps of PCR are shown in Figure 5.13 and are as follows:

- *Denaturation*: The double-stranded target DNA is heated to 92°C to separate (denature) the two strands by breaking the hydrogen bonds between the bases to produce single strands. Since there are three hydrogen bonds between guanine and cytosine, a DNA fragment with a high proportion of GC content would need a higher denaturation temperature.
- *Annealing*: In this step, the primers anneal or pair with the target DNA strands. The forward and reverse primers are synthesized according to the sequence of the target DNA. The annealing generally takes place at 50–60°C.

    Sometimes, primers may be allele-specific, namely, one primer is from a polymorphic region that is used in single nucleotide polymorphism (SNP). Degenerate primers are used for several target DNA. Linker primers possess restriction enzyme sites that are introduced in the target DNA for subsequent cloning.
- *Polymerization/elongation*: In this step, DNA polymerase extends or elongates the DNA using the dNTPs and DNA polymerase. This step is carried out at 72°C to maintain high fidelity. Since enzymes are heat labile, an enzyme isolated from the heat-resistant organism *Thermus aquaticus* is used.

FastStart polymerase is a variant of *Taq* polymerase that uses heat activation to avoid nonspecific amplification.

Pfu polymerase from *Pyrococcus furiosus* has proofreading ability to minimize errors. The polymerase from *Thermus thermophilus* has reverse transcriptase activity.

5.3.2.2 VARIATIONS OF PCR

- *Nested PCR*: In nested PCR, two sets of primers are used. The products of a first PCR reaction are used as templates for a second PCR where a different set of primers are used. These have annealing sites nested within the first set. Hence, this is used for specificity enhancement.
- *Real-time PCR*: It monitors the amplification of a targeted DNA molecule during the PCR, that is, in real-time, and not at its end, as in conventional PCR. When the PCR is used to quantify the number of template molecules, it is termed as quantitative real-time PCR, and as semiquantitative real-time PCR, to know the amount of DNA above/below a certain number. Real-time PCR uses fluorescent probes such as SYBR Green or fluorophore DNA probes.
- *Colony PCR*: Here, the bacterial colonies are transferred to the PCR mix. The DNA may be released by high heat treatment and is used as a template for amplification.
- *Inverse PCR*: This is for the amplification and sequencing of the surrounding DNA of a target. The target DNA is digested using restriction enzymes and self-ligated. The primers are extended outward from the target.
- *Suicide PCR*: Suicide PCR is used to eliminate nonspecific amplification. The primers should have been used only once, it ensures no contamination from previous PCR usage in the lab.

5.3.2.3 APPLICATIONS OF PCR

1. PCR is used in gene amplification and cloning for concentrating the target DNA that is available in very small quantities from any source.
2. It is used for DNA fingerprinting where we can identify one individual from a population. This has applications in forensics and paternity testing.

3. It can be used to amplify and study DNA from nanogram quantities in fossils having potential in paleontology.
4. It can be used for the diagnosis of pathogens such as HIV-1, hepatitis B, *Neisseria gonorrhea* and *Mycobacterium tuberculosis*.
5. It is used in diagnosing hereditary diseases.
6. It has application in epidemiological studies for the molecular characterization of pathogens.
7. It is an accurate and sensitive technique to detect the copy number of the gene or plasmids *in vivo*.
8. It is also used to introduce site-specific mutations in the gene of interest

5.3.3 DNA SEQUENCING

The determination of the nucleotide sequence in a precise order in the DNA is DNA sequencing. The techniques that are employed:

5.3.3.1 MAXIM AND GILBERT METHOD

It is also known as chemical sequencing, which employs radioactive labeling. The gene to be sequenced is first isolated and subjected to radioactive labeling by γP^{32}. This is followed by chemical treatment—purines (A and G) are treated with formic acid, guanine (G) with dimethyl sulfate, pyrimidines (C and T) to hydrazine, and NaCl and hydrazine for cytosine (C). Hence, we have four reactions—for A and G; G; T and C; and C. The products of these four treatments are separated on the base of size on agarose gel. Subsequent exposure to autoradiography will show the DNA bands. The sequence is then constructed on the basis of the patterns.[42]

5.3.3.2 SANGER SEQUENCING

This is based on chain termination by using dideoxynucleotides (ddNTPs). The template to be sequenced is to be in single-stranded form (a convenient way is to use M13 phage)—DNA primer, DNA polymerase, dNTP and ddNTPs—that terminate DNA elongation as they lack OH at the 3′

location preventing the formation of phosphodiester bonds. The ddNTPs can be radioactively/fluorescently labeled for their detection.[43]

The DNA is divided into four sets containing all the four dNTPs and polymerase. The four ddNTPs are added to one reaction each, that is, ddATP, ddCTP, ddTTP, and dGTP at lower concentrations. Following the extension, the products are separated on denaturing polyacrylamide gel based on size. Subsequent, autoradiography or fluorescent study in a sequencer reveals the DNA sequence to be constructed.

5.3.3.3   DYE TERMINATOR SEQUENCING

This method employs the use of ddNTPs that are labeled with fluorescent dyes that have different wavelengths at which they emit light. Hence, the four set reactions can be circumvented and a single test can be run. The use of this technique with automated sequencers has expedited the process of sequencing.

Other *next-generation high-throughput technologies* of sequencing have now been employed that encompass:

5.3.3.4   PYROSEQUENCING

Pyrosequencing entails the use of pyrophosphate release when nucleotides are incorporated in the DNA chain.[44] It is also called sequencing by synthesis. A single-stranded DNA template is immobilized and its complementary strand is synthesized using DNA polymerase coupled with a chemiluminescent enzyme (such as luciferase), by adding the four dNTPs, one at a time to allow base pair incorporation. When the correct base is incorporated by the polymerase, PPi is released. This is converted to ATP using ATP sulfurylase in the reaction mix. The ATP produced enables conversion of luciferin (present in the reaction mix) to oxyluciferin by luciferase producing visible light.

5.3.3.5   POLONY SEQUENCING

Polony sequencing combines an in vitro paired tag with PCR with high accuracy and low cost. Polony sequencing is generally performed on library with paired-end tags library with the length of each molecule of DNA template being 135 bp having two 17–18 bp paired genomic tags separated and flanked by common sequences.[45] The current read length of this technique is 26 bases per amplicon and 13 bases per tag, with each tag left with a gap of 4–5 bases. The protocol of polony sequencing can be broken into three main parts: paired end-tag library construction, template amplification, DNA sequencing, and finally sequence analysis.

5.3.3.6   SOLEXA (ILLUMINA) SEQUENCING

Illumina sequencing method has been developed by Solexa. Here, the DNA is amplified and extended one fluorescent labeled nucleotide at a time. It is based on the clonal amplification by polymerase of DNA on a surface of "DNA clusters or colonies."[46] For sequence determination, four types of reversible terminator bases are added and nonincorporated nucleotides are washed off. Images of the fluorescently labeled nucleotides are taken by a camera. This is followed by chemical removal of the dye, along with the terminal 3′ blocker, allowing the beginning of the next cycle. The DNA chains are extended one nucleotide at a time followed by the acquisition of an image.

5.3.3.7   DNA NANOBALL SEQUENCING

It is a high-throughput method for genome sequencing. DNA is amplified using the rolling model of replication.[47]

5.3.3.8   SINGLE MOLECULE REAL-TIME RNAP

This method uses RNA polymerase (RNAP) on an optical bead and sequenced DNA on another bead. The RNAP activity allows per nucleotide resolution.[48]

5.3.3.9   ION TORRENT SEQUENCING

This technique is based on using standard sequencing chemistry and detection by a semiconductor-based system of hydrogen ions that are released during the polymerization of DNA.[49] A single type of nucleotide is flooded in a microwell containing a template DNA strand to be sequenced. The nucleotide gets incorporated into the growing complementary strand if the introduced nucleotide is complementary to the leading template. This causes the release of a hydrogen ion that triggers a hypersensitive ion sensor, and corresponding number of released hydrogens and a proportionally higher electronic signal.

5.3.3.10   454 PYROSEQUENCING

The method amplifies DNA in emulsion PCR, that is, inside water droplets in an oil solution, with each droplet containing a single DNA template attached to a single primer-coated bead which forms a clonal colony. The sequencing machine contains many wells of picolitre volume, each containing a single bead and sequencing enzymes.[50] The light generated by luciferase is used for detection of the individual nucleotides added to the nascent DNA. Finally, sequence reads are generated by combining data.

5.3.3.11   APPLICATIONS OF DNA SEQUENCING

- The DNA sequence of a gene gives the amino acid sequence of the protein it encodes.
- The technique is used in DNA fingerprinting.
- The DNA sequences of entire genomes have been deduced and can be used in comparative gene analysis and phylogenetic and evolutionary biology.
- It is a powerful tool in the study of pathogens and other organisms to study molecular reasons and disease association.
- DNA sequencing with markers can be used by geneticists to develop transgenic plants and animals.

5.3.3.12   CASE STUDY

5.3.3.12.1   Expressed Sequence Tag

We very frequently come across this term in the process of sequencing of genes expressed in specific conditions. Sequencing of large genomes involves sequencing in parts of chromosomes, nonmethylated DNA and the expressed sequence tag (EST). EST is a short sequence of DNA resulting from one short stretch of sequence of a cloned cDNA.[51] ESTs are useful in sequence determination, gene discovery, and in identifying gene transcripts. Individual clones of cDNA library are used for EST generation. Nearly 300–800 nucleotides long fragments are generated, which represent portions of expressed genes. The ESTs are presented as either cDNA/mRNA sequence or as the reverse complement of the mRNA. The ESTs can be mapped to specific locations on a chromosome or can be aligned with an already sequenced genome of the organism that originated the EST.

ESTs are an important tool used to predict transcripts for genes, their protein products, and ultimately their function. Moreover, important information can be had of the external conditions from which the ESTs are obtained (such as specific tissue, organ, state of the organism—for example, developmental, environmental, spatial, etc.), and the obtained information is used to design precise probes for DNA microarrays to determine the gene expression. However, a major disadvantage of using ESTs is their high redundancy and the nonrepresentation of regulatory sequences.

Briefly, for the preparation of ESTs:

mRNA →cDNA → second strand cDNA synthesis → cleavage with restriction endonucleases → adaptor ligation → cloning → sequencing[37]

5.4 APPLICATIONS OF GENETIC ENGINEERING

5.4.1 MUTAGENESIS STUDIES

Modern molecular genetics requires mutations in organisms for several reasons. The main reasons are to understand molecular mechanisms underlying gene expression and gene regulation. For mutagenesis, the gene to be altered is carried on a plasmid, which is isolated, manipulated and returned to the host for expression of the gene. Mutations in genes can be carried out by generating deletions, random point mutations, and point mutations at a particular base pair.

5.4.2 GENE REGULATION STUDIES

Usually, a functional genetic unit (containing coding and regulatory sequences) is cloned in a vector and its sequence analysis is performed. This is followed by successive trimming to smaller sizes and re-cloning in a suitable plasmid. This clone is then transformed and the phenotype of the recipient cell is observed. When these regulatory sequences and structural genes are trimmed, a change in phenotype is observed. In *E. coli*, the lac operon was studied this way (see Chapter 3, Section 3.1.1).

5.4.3 INDUSTRIALLY IMPORTANT BACTERIA

Through genetic engineering, bacteria of industrially important phenotype can be produced. Several genes from different bacteria are inserted into a single plasmid (e.g., genes coding for antibiotic resistance, origin of replication, promoters, etc.). These assorted genes have been successfully placed in bacteria that can metabolize substrates in an environmental-friendly way; bacteria that can degrade waste more efficiently.[52] The use of these "modified" microbes play an important role in environment quality, world economy, and food production.

5.4.4 PLANT GENETIC ENGINEERING

An important application of recombinant DNA technology is genetic engineering of plants, in which genotypes of plants can be suitably altered. Genes from one plant can be cloned in a shuttle plasmid (e.g., Ti plasmid of *Agrobacterim tumefaciens)* and be transferred to another plant where the gene is intended to be expressed (Chapter 4, Section 4.3).

5.4.5 GENOME EDITING BY CRISPR AND ITS USE IN EUKARYOTIC SYSTEMS

The mechanism of clustered regularly interspaced short palindromic repeats (CRISPR) as a prokaryotic immune boosting system has been discussed in detail in Chapter 3 of this book. The Type II CRISPR system performs CRISPR/Cas9 genome editing. When utilized for genome editing, this system includes Cas9, CRISPR RNA (crRNA), transacti-vating crRNA (tracrRNA) along with an optional section of DNA repair template that is utilized in either nonhomologous end joining (NHEJ) or homology-directed repair (HDR). CRISPR genome-editing technology takes advantage of DNA repair to introduce specific mutations or DNA sequences into target genomes. Briefly, Crisper RNAs recognize their complementary targets and guide CRISPR-associated proteins (Cas9) proteins to target sequences. The Cas9 nuclease then introduces a double-strand break in the target DNA. This break can be repaired through two pathways: NHEJ or HDR as a result of which, either accurate repair or mutations are introduced.

The *NHEJ* may result in small insertions or deletions that can inacti-vate the target gene. By using two crRNAs, larger deletions can be gener-ated, and if sites on two different chromosomes are targeted, NHEJ can result in chromosomal translocations and inversions.

The *HDR* inserts pieces of foreign DNA by providing donor DNA. This can be used to generate conditional alleles, tag proteins, insert specific mutations, or correct existing mutations.[53] Single-strand breaks are prefer-entially generated by using a nickase mutant of Cas9 that cuts; this repair is biased toward HDR.

The CRISPR system can be used to activate or enhance transcription by fusing catalytically dead cas9 to a transcriptional activator and using a guide RNA that targets the promoter region of a gene. A gene can also be downregulated by fusing catalytically dead cas9 to a repressor. The catalytically dead cas9 when fused to a fluorescent reporter also serves as a molecular beacon, facilitating visualization of genomic loci of interest.

5.4.5.1   HOW CRISPR WORKS IN EUKARYOTES

Although CRISPR is a prokaryotic system, it has been adapted to work in almost any model organism.[54–58] This system (Type II CRISPR-Cas9; Chapter 3, Figs. 3.16 and 3.17; Fig. 5.14) has been the system of choice for most current genome-editing applications because it is the simplest system: it requires only a single protein, Cas9, for crRNA maturation, targeting, and target DNA cleavage, and an additional transactivating Crisper RNA for maturation and targeting. The transactivating Crisper RNA and Crisper RNA can be fused to form a single guide RNA that can be custom designed to target almost any sequence (Fig. 5.14; Chapter 3, Figs. 3.16 and 3.17).[54] This system can easily be used in both prokaryotes and eukaryotes by eliminating the need for the RNase III transactivating Crisper RNA: pre-Crisper RNA processing step. Gene editing in eukaryotes can thus be achieved simply by providing Cas9 and appropriate guide RNA.

For delivery of the Cas9 and gRNA into target cells, viral or nonviral systems can be used. Although the most common and cheapest system is electroporation, a more efficient delivery system is required for such as those based on lentivirus (LVs), adenovirus (AdV) and adeno-associated virus (AAV) in cells that are (stem cells, neurons, hematopoietic cells, etc.) difficult-to-transfect. Briefly, sgRNA and Cas9 can be co-expressed through a plasmid to be used for transfecting cells. The packaging together of multiple crRNA's and the tracrRNA forms a single-guide RNA (sgRNA). This sgRNA can be joined together with the Cas9 gene to form a plasmid that can be used for cell transfection. This is followed by transfection, expression of the plasmid and activation of Cas9. The crRNA helps the Cas9 protein to find the correct sequence in the host DNA and—depending on the Cas9 variant—creates a single or double strand break in the DNA.

5.4.5.2   APPLICATIONS AND ADVANTAGES OF CRISPR AS A GENOME-EDITING AND GENOME TARGETING TOOL

A wide range of CRISPR applications are emerging.[55] Several research groups have used CRISPR to introduce single point mutations (deletions or insertions) in a particular target gene, via a single guide RNA. Using a pair of guide RNA-directed Cas9 nucleases, large deletions or rearrangements in the genome, such as inversions or translocations can also potentially be induced. In addition, the catalytically dead cas9 version of the CRISPR/Cas9 system has been used to target protein domains for transcriptional regulation, epigenetic modification, and for visualizing specific genome loci microscopically.

CRISPR/Cas9 enables rapid probing of gene function at genome-scale by generating large gRNA libraries for genomic screening. Furthermore, in contrast to other genome-editing tools, including zinc finger and TALENs, where the redesign of the protein-DNA interface is required, the CRISPR/Cas9 system requires only redesign of the Crisper RNA to change target specificity.

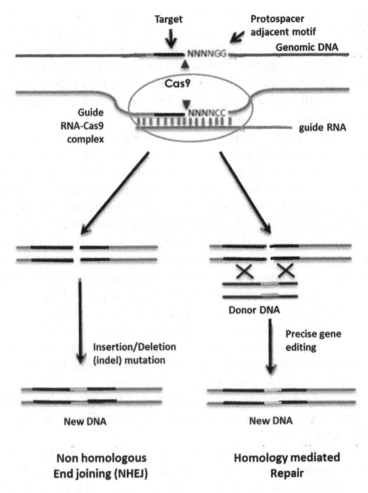

FIGURE 5.14 Guide RNA-Cas9 guided homologous and nonhomologous repair mechanisms. The guide RNA (gRNA) along with cas9 protein is introduced at sequences complementary to the target along in the cell and makes a double-stranded cut at 3–4 bases upstream of the protospacer adjacent motif (PAM) sequence (NGG) (shown as triangles). *Left panel*: NHEJ is an error-prone repair mechanism that does not use any template DNA to recombine but deletes or inserts (indel) random sequences in the cut to fill it up. This disrupts the original gene sequence. If a mutation in a gene needs to be rectified, a single- or double-stranded donor DNA with the corrected sequence flanking left and right homologous sequences of the gene is provided. *Right panel*: Due to the homology of the left and right flanking sequences, the donor DNA is used for homologous recombination by the cell to fill up the double-stranded cut in the DNA. The homology recombination will result in the exchange of DNA with the mutated sequence with the donor DNA containing the adjusted sequence resulting in the correction of mutation (right panel).

The CRISPR technique also makes it possible to correct mutations that result in human disease. CRISPR can be utilized to create cellular *models of diseases* affecting the humans. For instance, in human pluripotent stem cells, CRISPR was applied to introduce targeted mutations in genes relevant to polycystic kidney disease.[54,55] The CRISPR-modified pluripotent stem cells were subsequently grown into human kidney organoids that exhibited disease-specific phenotypes.[54,55]

The CRISPR/Cas9 system has been widely and successfully used to target important genes in many cell lines and organisms,[55] including *Drosophila*, human, *C. elegans*, bacteria, plants, zebrafish, *Xenopus tropicalis*, yeast, rabbits, monkeys, and pigs. The pig cells were engineered to inactivate all 62 porcine endogenous retrovirus in the pig genome, which eliminated infection from the pig to human cells in culture.[56]

Catalytically dead versions of Cas9 (dCas9) do not have CRISPR's DNA-cutting ability, but still possess the ability to target desirable sequences. Several regulatory factors have been added to dCas9s, enabling them to turn almost any gene on or off or subtly adjust its level of activity. dCas9 silence gene expression by base pairing with target gene without cutting the DNA, also known as "CRISPR interference" (CRISPRi).[57] The targeted site is methylated, thus modifying the gene. Thus, transcription is potentially inhibited by this modification. Silencing-specific genes can be effectively targeted and at the DNA level by Cas9.

As a novel form of antimicrobial therapy, CRISPR/Cas-based "RNA-guided nucleases" can be used to target microbial antibiotic resistance and virulence factors genes.

A few light or small molecules act as external triggers for gene activation or genome editing by several variants of CRISPR/Cas9. These *photoactivatable CRISPR systems*[58] have been developed by fusing an activator domain and a dCas9 for gene activation, with light-responsive protein partners or by fusing similar light responsive domains with two constructs of split-Cas9, or by modifying the guide RNAs with photo-cleavable complements for genome editing.[55,58]

5.4.5.3   THE FUTURE OF CRISPR/CAS9

A major advantage of CRISPR technology is that it relies on endogenous cellular mechanisms to silence or edit genes. Therefore, technically gene editing under the CRISPR method is not classified as a genomic modification as no new DNA is integrated in the host genome. With the potential of CRISPR/Cas9 reaching beyond simple DNA cleavage, its usefulness for genome locus-specific recruitment of proteins is highly remarkable.

Due to its simplicity, high efficiency, and versatility in CRISPR/Cas9 technology, there has been significant progress in developing Cas9 into a set of tools for cell and molecular biology research. Moreover, the CRISPR/Cas system seems to be the most user-friendly in comparison to other nuclease systems available for genetic engineering.

5.4.6 *PROTEIN EXPRESSED THROUGH RECOMBINANT DNA TECHNOLOGY*

The development of technologies for nucleic acid manipulation in the 1970s led to the construction of recombinant DNA molecules consisting of nucleotide sequences of different sources. This process is known as gene cloning, and the process of manipulation of genes is genetic engineering.

The initial step in recombinant technology begins with the isolation of a gene of interest (target gene) and insertion into a suitable vector to form replicon. The replicon is then introduced into a host to be cloned. The cloned replicon is the recombinant DNA. Cloning produces several copies of the recombinant DNA since the initial supply may be insufficient for use. A protein that is encoded by a recombinant DNA consisting of gene encoding the protein in a suitable vector that can express the gene and translate its messenger RNA is called a recombinant protein. The basic steps of recombinant DNA technology involve:

- bioinformatic analysis, followed by isolation/amplification of target gene and a suitable vector,
- cutting/cleavage of the target gene and vector so that they have compatible ends to be joined,
- joining/ligation of the target gene and vector,
- transforming and cloning,
- selection/screening of the desired clone,
- expression of the clone in a suitable host, and
- detection and purification of the recombinant protein.

5.4.6.1   RECOMBINANT PROTEINS

A few challenges such as codon bias, protein conformation, insolubility, stability, purification yields, and host cell toxicity should be resolved when microbial hosts are used to express recombinant proteins.

The use of codons for amino acids ranges from one (methionine and tryptophan) to as many as six codons (arginine, leucine, and serine) in the standard genetic code (see Chapter 2). Different organisms use codons with varying preferences, also known as *codon bias*. Thus, if only one possible DNA sequence encodes a specific protein, it puts a limitation in designing criteria such as elimination or incorporation of restrictions enzyme sites and sequences that can compromise stability or form structures at or around the translational initiation site. Therefore, those codons can be used whose frequency of use matches that of the host codon in the designing gene. The effect of codon bias is also observed in yeast, plant, fungal, and mammalian hosts.[59,60]

To address these challenges of production and purification efficiency, fusion tags are incorporated to increase expression yields and influence solubility and native folding.[61] A number of fusion proteins have been used for this purpose, including, glutathione S-transferase (GST), maltose-binding protein (MBP), thioredoxin A (TrxA), small ubiquitin-related modifier (SUMO), and ketosteroid isomerase (KSI).

A short sequence (i.e., an epitope) is added to a protein of interest by recombinant DNA methods by a technique known as *epitope tagging*. Epitopes enable post-translation modifications, protein interactions, intracellular trafficking, and determination of protein size and concentration. Affinity tagging is used in a variety of applications including western blot analysis, immunoprecipitation (IP), co-immunoprecipitation (co-IP), immunofluorescence (IF), and affinity purification. FLAG, c-myc, HA, 1D4, polyArg and polyHis, and Streptavidin binding tags are a few widely used affinity tags.[61]

When the presence of the fusion tag affects the structure or biological function of the protein of interest, it becomes essential to remove carrier proteins and affinity tags. For this, site-specific proteases cleave specific sequences that have been included between the tag(s) and native protein. Common proteases include enterokinase, factor Xa, SUMO protease, tobacco etch virus (TEV) protease, thrombin, and 3C.[62]

5.4.6.2   RECOMBINANT PROTEIN EXPRESSION SYSTEMS

The recombinant proteins have been successfully expressed in both prokaryotic and eukaryotic systems. *E. coli, Bacillus subtilis, Staphylococcus carnosus, Streptomyces lividans* are a few prokaryotic systems that are routinely used for recombinant protein expression. The presence of prokaryotic promoter, ribosome binding site, multiple cloning sites, transcription termination site as well as a selectable marker is a must for successful expression of recombinant proteins. Whereas faster cell growth, simple medium and lesser cost of production are major advantages, degradation by bacterial proteases, improper folding and biofilm formation are a few problems associated with using prokaryotic expression systems.

Likewise, yeast, *Aspergillus niger*, baculovirus—insect cells, mammalian cells (e.g., Chinese hamster ovary cells) are a few recombinant protein eukaryotic expression systems. The expression vectors must possess eukaryotic promoter, multiple cloning sites, eukaryotic transcription termination site, and a selectable marker. Unlike the prokaryotic expression systems, eukaryotic expression systems allow for posttranslational modification (PTM) of proteins (e.g., glycosylation, phosphorylation, hydroxylation, disulfide bond formation, and processing of propeptides).

5.4.6.3   FEW RECOMBINANT PROTEINS WITH POTENTIAL IN HUMAN LIFE

5.4.6.3.1   Antibodies

These are Y-shaped glycoproteins produced by the B-lymphocytes in response to an antigen (foreign particle). Antigens may possess various epitopes that are recognized by many lymphocytes. When several lymphocytes elicit antibody production against an antigen, it is *polyclonal*.

Monoclonal antibodies are produced by a single B lymphocyte clone that is specific for one epitope of an antigen. In 1970s, Kohler and Milstein fused B cells of spleen with myeloma cells to produce hybridomas that produced monoclonal antibodies and were immortal.[63]

The use of antibodies in therapy is growing for treatment in various diseases such as cancer, autoimmune responses, etc. Rather than producing antibodies from cell lines, the use of transgenic animals holds promise. The gene for the human antibody against a pathogen can be introduced into the germ line of an animal. The animal would produce antibodies upon exposure to the pathogen and can be used as a "bioreactor."

Fermentation in *E. coli* at high-cell density can yield up to 1–2 g/L depending on the individual antibody fragment.[64] Antibody fragments expressed in *E. coli* are mainly secreted into the periplasmic space and need to be extracted from there. However, larger antibody formats are very difficult to express in bacteria; moreover, the lack of a glycosylation apparatus limits their use, if effector functions are needed.

Compared to prokaryotes, insect cells contain a better protein folding mechanisms and secretion apparatus. However, the current scenario demands even more developed stable insect cell lines and process technology. As a result, mammalian cell lines are most widely used and preferred for therapeutic antibody production as they provide folding, glycosylation, and secretion apparatus that are suitable as well as and more humanlike. However, for obtaining high levels of recombinant antibodies, high technical efforts are needed which are relatively economically burdensome. While recombinant proteins expression and production have been tried in several livestock animal species, generation of transgenic animals also is very laborious. An interesting approach is the hybridoma technology which combines with human transgenic animals for the developing human antibodies.

5.4.6.3.2 Vaccines

A vaccine is a preparation of an antigen that is injected in a target organism to elicit the production of an antibody response and hence, provide immunity to the antigen. Traditionally, the pathogens producing toxins are cultured artificially and then toxins are harvested and manipulated into toxoids. These toxoids or inactivated antigens can be mixed with an adjuvant (a substance that enhances antibody response such as alum) and administered to an organism.

Different pathogenic genes with use as vaccines have been expressed in hosts (Nascimento and Leite, 2012). An example is a recombinant vaccine against human papillomavirus (HPV)-recombinant proteins have been produced in insects and yeast.[66]

Several viral vectors such as vaccinia, AdV, alphavirus, etc. are available as vectors. Antigens of HIV have been expressed in AdVs vectors with promising results in phase-I studies.

DNA vaccines: DNA vaccines employ the use of a naked DNA as a vaccine. It is also called genetic vaccine. These vaccines show great promise, and several types are being tested in humans. DNA vaccines use the genes that code for those all-important antigens. It is a plasmid with an origin of replication, a cytomegalovirus promoter, multicloning site, and a selection marker (antibiotic). The DNA vaccines enable the host to present the antigenic protein similar to an infection and elicit an appropriate antibody response including the presentation by major histocompatibility complex (MHC) molecules. DNA vaccines are administered intramuscularly or by employing a nasal spray or by using a gene gun or biolistic delivery.

DNA vaccines have several advantages over other types of vaccines. Although purification of antigens is not required for DNA vaccines, they elicit higher immune response compared to other kinds of vaccines. Moreover, these vaccines trigger immune response only against the specific pathogen because of their specificity in producing target proteins. Their storage and transport concerns are minimal since they maintain their stability at variable temperatures better than conventional vaccines.

The advantages of a DNA vaccine are that they induce humoral and cell-mediated responses and overcome issues plaguing recombinant protein production. Antigens from HIV, malaria, TB have been employed as DNA vaccines and have induced responses in animals.

A major disadvantage of using DNA vaccines is that the DNA could possibly get integrated into the host genome. In addition, the PTM of the gene (DNA vaccine) product in host cells may not be the same as that found in the native antigen.

Edible vaccine: The edible vaccines can be easily ingested by eating edible plant parts. This eliminates the laborious and costly processing and purification procedures that are otherwise mandatory. Transgenic plants (e.g., tomato and potato) have been developed for expressing antigens derived from animal viruses (rabies and herpes viruses). The production of vaccine in potatoes is illustrated in Fig. 4.16, Chapter 4.

Attenuated recombinant vaccines: Attenuated recombinant vaccines are genetically engineered (bacteria or viruses) that are live vaccines.[65–67]

Broadly two types of genetic manipulations are carried out for the production of these vaccines:

1. The virulence genes of pathogenic organisms are deleted or modified.
2. Nonpathogenic organisms are genetically manipulated to carry and express antigen determinants from pathogenic organisms.

The native conformation of immunogenic determinants of attenuated vaccines is preserved thus substantially increasing the immune response. Some of the attenuated vaccines developed by genetic manipulations are described briefly in the coming sections.

By genetic engineering, it was possible to create a new, nonpathogenic strain of *Vibrio cholera* by deleting the DNA sequence encoding the A$_1$ peptide of *V. cholera* enterotoxin, a toxin causing symptoms of cholera in a host. It is a hexamer consisting of one A subunit and five B subunits. The new strain cannot produce enterotoxin and is a good attenuated vaccine candidate.

5.4.6.4   CASE STUDY

5.4.6.4.1   *Vector Recombinant Vaccines*

Some of the vectors can be genetically modified (GM) and employed as vaccines against pathogens.

Vaccines against vaccinia virus: The genome of vaccinia virus can accommodate stretches of foreign DNA which can be expressed along with the viral genes.[68] The vaccinia viruses are generally harmless, relatively easy to cultivate and stable for years after lyophilizing (freeze-drying). All these features make the vaccinia virus strong candidates for being used as vector vaccine. For this, the foreign genes (from a pathogenic organism) can be inserted into vaccinia virus genome for encoding antigens; this, in turn, produces antibodies against the specific disease-causing agent.

The vaccinia virus can replicate in the cytoplasm of the infected cells rather than the nucleus because it possesses its own machinery for DNA replication, transcription-DNA polymerase, RNAP, etc. As a result, foreign genes inserted into the vaccinia virus can also be expressed along with the viral independent of the host cell genome.

The vector vaccine can stimulate both B- and T-lymphocytes, in contrast to a subunit vaccine which can stimulate only B-lymphocytes. Thus, a high level of immunoprotection against pathogens is acquired. A recombinant vaccinia virus which carries genes encoding different antigens can be used for vaccinating individuals against different diseases simultaneously.

Although vector vaccines have been developed against hepatitis, influenza, herpes simplex virus, rabies, angular stomatitis virus, and malaria, none of these vaccines has been licensed for human use due to fear of safety. The most important limitation is the yet unknown risks of using these vaccines in humans.

Delivery of antigens by bacteria: The observation that the antigens located on the surface of a bacterial cell are more immunogenic than the antigens in the cytoplasm,[69] scientists have developed strategies to express the surfaces of organisms that are non-pathogenic with antigens that are originally expressed in pathogenic bacteria. For example, flagellin is a protein present in the flagella, the threadlike filaments, of *Salmonella*. A synthetic oligonucleotide encoding the epitope of cholera toxin B subunit was inserted into *Salmonella* flagellin gene and was found to express on the flagellum surface. These flagella-engineered bacteria, when administered to mice, raised antibodies against the cholera toxin B subunit peptide. Thus, in future, it may also be potentially possible to incorporate multiple epitopes (two or three) into the flagellin gene to create multivalent bacterial vaccines.

5.4.6.4.2 *Recombinant (Protein Subunit) Vaccines*

They are also called subunit vaccines and containing only a fraction of the pathogenic organism. Often, these are wither synthetic peptides that represent the protein component which induces an immune response,[70] they can also consist of protein subunits (antigens) expressed in heterologous systems (*E. coli*, yeast, insect, etc.) using recombinant technologies. Currently, several vaccines under investigation are derived from such purified recombinant proteins or subunits of antigens. One of the best examples of recombinant protein vaccine currently in use in humans is the vaccine against hepatitis B virus (HBV), which has been schematically depicted in Figure 5.15.

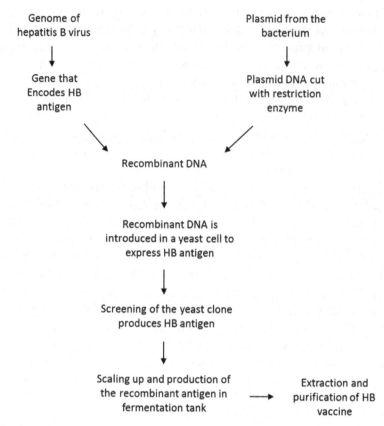

FIGURE 5.15   Flowchart showing the production of recombinant HB vaccine.

5.4.6.5   PRODUCTION OF RECOMBINANT THERAPEUTICS

Various proteins and products of living cells such as cytokines, growth factors, hormones, and other regulatory peptides and proteins can be used in the treatment and prevention of diseases. They were often originally extracted from human tissues and secretions. With the advent of more sophisticated technologies, it has become possible to prepare pure, highly concentrated materials. The diseases that can be treated are Crohn's disease, rheumatoid arthritis, multiple sclerosis, diabetes, etc.

The simplest system for production of recombinant proteins is by using *E. coli*, which has the advantage of being well-studied organism and has many vectors available. This organism grows fast in a culture medium,

produces a large quantity of products, is relatively inexpensive to cultivate, and multiplies rapidly.

The range of recombinant medicinal products on the pharmaceutical
market includes:

Growth factors: erythropoietin, thrombopoietin, platelet growth
factors, bone morphogens. Erythropoietin is a hormone that stimulates the
stem cells of bone marrow to produce mature erythrocytes. It is synthesized by the kidneys.

Cytokines (e.g., interferon alpha, granulocyte colony-stimulating
factor). Interferons are used for the treatment of a large number of viral
diseases and cancers such as leukaemia, Kaposi's sarcoma, bladder cancer,
head and neck cancer, renal cell carcinoma, skin cancer, and multiple
myeloma.

Hormones: insulin, glucagon, human growth hormone, follicle stimulating hormone, luteinizing hormone, human chorionic gonadotropin. The
human growth hormone gene is cloned into *E. coli* and purified. Production of recombinant insulin hormone regulates the level of glucose in
blood. People who produce less amounts or dysfunctional proteins suffer
from diabetes mellitus. For many years, patients were administered insulin
that was extracted and processed from the pancreas of cows or pigs due to
the similarity in insulin structures.

With the advent of rDNA technology, it was possible to insert human
insulin gene in *E. coli* and express it. The genes for insulin A chain and B
chain are separately inserted to the plasmids of two different *E. coli* cultures.
The *lac* operon system has been used for expression of both the genes. The
presence of lactose in the culture medium, in separate cultures, induces the
synthesis of insulin α and β chains. The so formed insulin chains can be
isolated, purified, and joined together to give an entire protein.[71] Attempts
have been made in recent years to produce second-generation insulin by
site-directed mutagenesis and protein engineering. The second-generation
recombinant proteins are termed as muteins. Among these is insulin lispro,
with modified amino acid residues of the insulin β chain.

Protease enzyme: Tissue plasminogen activator (tPA) is a naturally
occurring protease enzyme that helps to dissolve blood clots. tPA was the
first pharmaceutical product to be produced by mammalian cell culture.

Coagulating proteins such as factors VIII and IX: The cDNA of the
mRNA of Factor VIII has been synthesized and cloned into mammalian
cells or hamster kidney cells for the production of recombinant factor VIII.

Enzymes, such as DNAse: This enzyme is very useful in the treatment of common hereditary disease cystic fibrosis. The biotechnology firm Genentech isolated and expressed the gene to produce recombinant DNase I and was found to clear the bacterial DNA accumulating in the bronchi.

Genetically engineered crops have one or few foreign genes into the best-accepted cultivar background. It must be noted that "GM" crops and foods are not new. For instance, due to extensive selective molecular farming practices (see Chapter 4, Fig. 4.6), members of the same genus or species of corn (*Zea mays*) or soybeans (*Glycine max*) are a result of human directed genetic modification of the original wild types. Modern bread wheat is a mixture of recombined genes from three different wild species and the strawberry you buy in stores is a mixture of genes from two species.

A major difference between conventional plant breeding techniques and genetic engineering is that potentially negative genes could be introduced in order to obtain one gene with the desirable trait. While induced mutagenesis has been used for decades to create genetic variants, the main advantage of genetic engineering is bridging the species barriers. The use of GM crops has created new opportunities—with new safety guidelines.

Some GE plants in field trials are[72]: maize, wheat, barley, rice, eggplant, strawberry, grape, potato, tomato, cucumber, sweet potato, and cabbage; oil-producing crops such as oilseed rape, flax, sunflower, mustard, soybean, peanut, and eucalyptus; and ornamental plants such as chrysanthemum, petunia, carnation, and cotton.

5.5   BIOTECHNOLOGY AND ETHICS

However, similar to all new technologies there are concerns with the commercial application GM technology and use of GM products.

5.5.1   *ENVIRONMENTAL IMPACT*

5.5.1.1   *IMPACT ON NONTARGET ORGANISMS*

The impact of GM crops on nontarget organisms should be considered. Ideally, a pest control mechanism should control the pest, and not harm other nontarget organisms. For example, the nontarget organisms could be birds, fish, mammals, etc.

The natural enemies of our crop pests are nontarget organisms and their proliferation can be encouraged. Various parasitic and predatory insects and other arthropods constitute the natural enemies. Control of crop pests by natural enemies is referred to as biological control. GM crops produce certain toxins that are more selective than pesticide sprays. These toxins do not harm natural pest predators. Thus, GM crops that produce their own plant pesticides are more compatible with biological control.

5.5.1.2   FOOD SAFETY

Extensive studies indicate that GM foods are as safe as conventional foods. In addition, strict regulatory oversight and more extensive safety standards for biotech foods such as mandatory labelling of the presence of common allergens in certain foods and scientific proof that allergens are not present in the modified food. Moreover, the product must be labeled accordingly if there is any change in nutritional content or composition.

5.5.2   SOCIOECONOMIC CONSIDERATIONS

The major social and economic concerns that have been raised are: the biotechnology benefits larger group rather than the wealthier farmers and larger corporations; the control of intellectual property associated with agricultural products and the associated benefits, in other words, does it promote only monetary benefits; the possibility of increased reliance of farmers on seed companies and less likely to use seeds of traditional varieties.

5.5.3   ETHICAL CONCERNS

The socioeconomic issues can also be ethical considerations. Because the adoption of biotechnology in some ways is nontraditional, some of the major ethical concerns that have been raised are: to what extent do we manipulate the nature; can the genetically engineered plants and animals be considered natural; should the patenting or ownership of a GM product be allowed.

Intense debates have been raised regarding the safety of genetic modification and hence, its potential adverse or unpredictable effects in nature. In the context of biotechnology, "naturalness" is a much-debated topic.

One of the major fears of genetic engineering is the *behavior of GM organisms*. Regarding the use of GM microbes used as vectors—their inadvertent or purposeful release in nature may have unforeseen or unpredicted effects in an ecological niche.

Another concern is if these GM organisms interact or breed with the wild type population, there could be *mixing of the gene pools*. A fear that stems from this belief is the displacement or contamination of the wildtype.

A further issue is the *horizontal transfer of genes*. This involves the transfer of DNA across different genomes. It occurs through recombination in prokaryotes and is a complex mechanism in eukaryotes. Prokaryotes can exchange DNA with eukaryotes, through the probable mechanisms of conjugation and endocytosis. The issue can be addressed by looking at the phenomenon of evolution. However, the emergent picture is that horizontal gene transfer plays a larger role in microbial evolution than previously thought.[73] The horizontal gene transfer has imparted certain survival strategies naturally in the course of time.

To overcome these controversies, the development and use of such GM organisms would need regulation by a legislative body. The *Genetic Engineering Approval Committee* (*GEAC*) has been installed and vested with the powers to regulate the genetic modification activity.

Apart from these issues, there are *moral and ethical* issues regarding the use and modification of life forms, for example, the use of transgenic animals such as nude mice have evoked pathos and condemnation. But, the moral values differ widely across nations and cultures. These beliefs are not permanent and are bound to change with the passage of time.[74]

Another point that can be noted is that the pathos evoked may vary with the type or species of life form used. For example, people may feel bad when mice or dogs are used in transgenics but the same may not apply when we discuss certain insect forms such as locusts. So, the ethical dilemma here is subject to an individual's personal outlook and prejudices. As here, the point that certain animals are routinely reared and consumed as food by slaughter so these points have to be taken into consideration before condemning transgenics.

Religious beliefs do have significant hold as certain objections to biotechnology is the "act of playing God" by the creation or alteration of life forms genetically. The main goal of biotechnology "is the integration of organisms, cells parts and thereof molecular analogues for products and services" (European Federation of Biotechnology definition). This issue is again an intrinsic perspective that relies on an individual's outlook on "what is natural?"

A further concern is the use of products that have been produced in an organism that is sacred or condemned by certain religious outlooks. But, if the biotechnological product in question would help to alleviate disease or suffering or even prolong the extent and quality of life, then these concerns can be safely circumvented by the belief in the picture.

The above two points mainly ethical/moral and religious beliefs are complex and are subject to the variables of time.

5.5.4   CASE STUDIES

Golden rice is a product of modified to contain β carotene: a precursor of vitamin A. The safety and efficacy of golden rice and its utility in humans as a source of vitamin A has been established. Despite its merits, the release and consumption was opposed by activists citing safety concerns.

Transgenic plants expressing the toxin of *Bacillus thuringenesis:* for example, Bt cotton. These plants have been transformed with the *cry* gene of the soil bacterium using *Agrobacterium tumefaciens.* These transgenic plants express a toxin that confers resistance against bollworm infection that causes severe losses to crops. Despite its promising results, it is yet to see light in several countries including India.

Biowars: Deliberate construction and use of harmful biological agents with potential serious hazard could be a cause of biowars. Biological Weapons Conventions of 1972 has been signed by most countries that extracts a promise of no usage or production of such acts.

Sequence information from human genome: The growth in the use of high-throughput sequencing of human genome has introduced a number of ethical issues. The ownership of an individual's DNA and the data produced when that DNA is sequenced is an important issue. Moore v. Regents of the University of California: in 1990 ruled that "individuals have no property rights to discarded cells or any profits made using these cells (for instance, as a patented cell line).[75] However, individuals have a right to informed consent regarding removal and use of cells," Regarding the data produced through DNA sequencing, *Moore* gives the individual no rights to the information derived from their DNA.

As DNA sequencing becomes more widespread, the issues of storage, security, and sharing of genomic data have become important and need to be dealt with. The Genetic Information Non-discrimination Act (GINA) prohibits discrimination on the basis of genetic information with respect to health insurance and employment. But particularly sensitive is the whole-genome sequencing data, as it could be used to identify not only the source individual but also their relatives. Increased use of genetic variation screening, both in new-borns and in adults have led to several ethical issues. The screening for genetic variations can be harmful, increasing anxiety in individuals who find out that they have an increased risk of a particular disease.

Editing human embryos: Human embryos have been edited by CRISPR.[76] Unlike somatic mutations, mutations made in the germ line are present in all cells of our body and passed on to the progeny, thus integrating the mutations permanently into collective human genome. But the ease of use, affordability, and differences in human germ line editing regulations in different countries, make it extremely challenging to ensure that correcting deleterious mutations will be CRISPR's only use.[77] Another concern is whether we truly understand biology well enough to intervene in such a drastic way.

In conclusion, the issue to debate is whether we should play with our own evolutionary trend and bend it to what we believe will lead to a better or healthier human being? By intervening in our genetic evolution, we are probably reducing diversity and creating genetic bottlenecks.

5.6   SUMMARY

Genetic engineering involves the manipulation of genes of an organism to express new desired traits. Examples of genetically engineered organisms include bacteria, plants, yeast, and animals. Genetic manipulation is done by physically moving a gene into a recipient organism and expressing the gene. This process of isolating the desired gene from the organism and expressing them through a plasmid in a host is called cloning and heterologous gene expression to produce recombinant therapeutic proteins, hormones, drugs, genetically engineered bacteria for breakdown of waste products and oil slicks, transgenic plants to produce vaccines, drugs, and other desired proteins. However, the use and production of these products and genetically engineered organisms is a topic of intense debate and needs stringent checks before being used.

5.7   QUESTIONS

1. List and briefly explain the important considerations when deciding on cloning and expressing a gene in a heterologous host system.
2. How would you remove one of the sticky ends if a gene has been cloned using *BamH*I site? How would you alter the sequence of one of the *BamH*I ends for this purpose? (Hint: you could consider partial digestions or another specific enzyme).
3. What is the preferred way of stabilizing a PCR fragment? What are the additional advantages of using those vectors? (Hint: you could consider using certain specific vectors.)
4. a. Describe the advantages of using bacteria host for producing heterologous proteins.
    b. How would you choose a vector in relation to genetic engineering?
    c. Define a plasmid in relation to genetic engineering?
5. Describe the following terms:
    a. Target gene
    b. Shuttle vector
    c. Restriction enzyme
    d. DNA ligase
    e. Expression library
6. Do you agree that genetic engineering should be allowed to make people more intelligent or make "designer babies?"

KEYWORDS

- **genetic engineering techniques**
- **cloning vectors**
- **biotechnology and ethics**
- **hybridization techniques**
- **PCR**
- **sequencing**
- **recombinant DNA**

REFERENCES

1. Berg, P.; Mertz, J. E. Personal Reflections on the Origins and Emergence of Recombinant DNA Technology. *Genetics* **2010**, *184*(1), 9–17.
2. Cohen, S.; Chang, A.; Boyer, H.; Helling, R. Construction of Biologically Functional Bacterial Plasmids In Vitro. *Proc. Natl. Acad. Sci. USA.* **1973**, *70*(11), 3240–3244.
3. Cohen, S. N. DNA Cloning: A Personal View After 40 Years. *Proc. Natl. Acad. Sci.* **2013**, *110*(39), 15521–15529.
4. Morange, M.; Cobb, M. *A History of Molecular Biology*; 1st ed.; Harvard University Press: Cambridge (MA), 2000.
5. Pingoud, A.; Jeltsch, A. Structure and Function of Type II Restriction Endonucleases. *Nucl. Acids. Res.* **2001**, *29*(18), 3705–3727.
6. Murray, N. E. Type I Restriction Systems: Sophisticated Molecular Machines (a Legacy of Bertani and Weigle). *Microbiol. Mol. Biol. Rev.* **2000**, *64*(2), 412–34.
7. Szebereny, J. Two Restriction Endonucleases. *Biochem. Mol. Biol. Edu.* **2006**, *34*(3), 228–229.
8. Lehman, I. R. DNA Ligase: Structure, Mechanism, and Function. *Science* **1974**, *186*(4166), 790–797.
9. Wilson, R. H.; Morton, S. K.; Deiderick, H.; Gerth, M. L.; Paul, H. A.; Gerber, I.; Patel, A.; Ellington, A. D.; Hunicke-Smith, S. P.; Patrick, W. M. Engineered DNA Ligases with Improved Activities In Vitro. *Protein Eng. Des. Sel.* **2013**, *26*(7), 471–8. DOI:10.1093/protein/gzt024. PMID 23754529.
10. Coleman, J. E. Structure and Mechanism of Alkaline Phosphatase. *Annu. Rev. Bioph. Biom. Str.* **1992**, *21*, 441–483.
11. Lodish, H.; Berk, A.; Zipursky, S. L., et al. *Molecular Cell Biology*; 4th ed. W. H. Freeman: New York, 2000. (Section 7.1, DNA Cloning with Plasmid Vectors.) https://www.ncbi.nlm.nih.gov/books/NBK21498/
12. Junwen Mao; Ting Lu. Population-dynamic Modeling of Bacterial Horizontal Gene Transfer by Natural Transformation. *Biophys. J.* **2016**, *110*(1), 258–268.
13. Smith, L. C.; Bordignon, V.; Babkine, M.; Fecteau, G.; Keefer, C. Benefits and Problems with Cloning Animals. *Can. Vet. J.* **2000**, *41*(12), 919–924.
14. Phil, T.; Alexander, M.; Andy, B.; Mike, W. In *Ligation, Transformation and Analysis of Recombinants, Design of Plasmid Vectors*. Instant Notes, *Molecular Biology*; Taylor and Francis Group: London, UK: 2005; pp 119–126.
15. Allison, L. A. *Fundamental Molecular Biology;* 1st ed.; Wiley-Blackwell: Malden (MA), 2007.
16. Ausubel, F. M.; Brent, R.; Kingston, R. E.; Moore, D. D.; Seidman, J. G.; Smith, J. A.; Struhl, K., eds.; Short Protocols in Molecular Biology; 5th ed.; John Wiley & Sons: New York, 2002.
17. Sambrook, J., Russell, D. W. *Molecular Cloning: A Laboratory Manual;* 3rd ed.; Cold Spring Harbor Laboratory Press: New York, 2001.
18. Langley et al. Molecular Basis of Beta β-galactosidase Structural Gene of *E. coli. J. Mol. biol.* **1975**, *24*(2), 339–343.
19. Yanisch-Perron, C.; Vieira, J.; Messing, J. Improved M13 Phage Cloning Vectors and Host Strains: Nucleotide Sequences of the M13mp18 and pUC19 Vectors. *Gene* **1985**, *33*(1), 103–119.

20. Lodish, H.; Berk, A.; Zipursky, SL., et al. *Molecular Cell Biology.* 4th ed. W. H. Freeman: New York, 2000. (Section 7.2, Constructing DNA Libraries with λ Phage and Other Cloning Vectors.) https://www.ncbi.nlm.nih.gov/books/NBK21696/.

21. Hedges, R. W.; Jacob, A. E. Transposition of Ampicillin Resistance from RP4 to Other Replicons. *Mol. Gen. Genet.* **1974,** *132,* 31–40.

22. Szmulewicz, M. N.; Novick, G. E.; Herrera, R. J. Effects of Alu Insertions on Gene Function. *Electrophoresis* **1998,** *19*(8–9), 1260–1264. DOI:10.1002/elps.1150190806. ISSN 0173-0835.

23. Muñoz-López, M.; García-Pérez, J. L. DNA Transposons: Nature and Applications in Genomics. *Curr. Genomics* **2010,** *11*(2), 115–128.

24. Ramsay. Yeast Artificial Chromosome Cloning. *Mol Biotech.* **1994,** *1*(2), 181–201.

25. Shizuya, H.; Hosein, K-M. The Development AND Applications of the Bacterial Artificial Chromosome Cloning System. *Keio. J. Med.* **2001,** *50*(1), 26–30.

26. Kuroiwa, Y.; Kasinathan, P.; Choi, Y. J.; Naeem, R.; Tomizuka, K.; Sullivan, E. J.; Knott, J. G.; Duteau, A.; Goldsby, R. A.; Osborne, B. A., et al. Cloned Transchromosomic Calves Producing Human Immunoglobulin. *Nat Biotechnol.* **2002,** *20,* 889–894.

27. Kozak, M. Point Mutations Define a Sequence Flanking the AUG Initiator Codon That Modulates Translation by Eukaryotic Ribosomes. *Cell* **1996,** *44,* 283–292.

28. Thieme, F.; Engler, C.; Kandzia, R.; Marillonnet, S. Quick and Clean Cloning: A Ligation-independent Cloning Strategy for Selective Cloning of Specific PCR Products from Non-specific Mixes. *PLoS One* **2011,** *6*(6): e20556. DOI:10.1371/journal. pone.0020556.

29. Petersen, L. K.; Stowers, R. S. A Gateway MultiSite Recombination Cloning Toolkit. *PLoS ONE* **2011,** *6*(9): e24531. DOI:10.1371/journal.pone.0024531.

30. Katzen, F. Gateway(®) Recombinational Cloning: A Biological Operating System. *Expert. Opin. Drug. Discov.* **2007,** *2*(4), 571–89.

31. Gibson, D. G. Enzymatic Assembly of Overlapping DNA Fragments. *Methods Enzymol.* **2011,** *498,* 349–361.

32. Holton, T. A.; Graham, M. W. A Simple and Efficient Method for Direct Cloning of PCR Products Using ddT-tailed Vectors. *Nucleic Acids Res.* **1991,** *19*(5), 1156.

33. Engler, C.; Gruetzner, R.; Kandzia, R.; Marillonnet, S. Golden Gate Shuffling: A One-pot DNA Shuffling Method Based on Type IIS Restriction Enzymes. *PLoS One* **2009,** *4,* e5553.

34. Raymond, C. K.; Olson, M. V.; Sims, E. H. Linker-mediated Recombinational Subcloning of Large DNA Fragments Using Yeast. *Genome Res.* **2002,** *12,* 190–197.

35. Jang, Y.-S.; Jung, Y. R.; Lee, S. Y.; Kim, J. M.; Lee, J. W.; Oh, D.-B., ... Kang, K. Y. Construction and Characterization of Shuttle Vectors for Succinic Acid-producing Rumen Bacteria. *Appl. Environ. Microbiol.* **2007,** *73*(17), 5411–5420.

36. Frazer, L. N.; O'Keefe, R. T. A New Series of Yeast Shuttle Vectors for the Recovery and Identification of Multiple Plasmids from *Saccharomyces cerevisiae. Yeast* **2007,** *24*(9):777–789.

37. Daniel, P.; Tomkins, J.; Frisch, D. Construction of Plant Bacterial Artificial Chromosome (BAC) Libraries: An Illustrated Guide. *J. Agricult. Gen.* **2000,** *5,* 1–100.

38. Davis, C. A.; Benzer, S. Generation of cDNA Expression Libraries Enriched for In-frame Sequences. *Proc. Natl. Acad. Sci. USA*. (In Applied Biological Sciences). **1997,** *94,* 2128–2132.

39. Southern, E. M. Detection of Specific Sequences Among DNA Fragments Separated by Gel Electrophoresis. *J. Mol. Biol.* **1975,** *98*(3), 503–517.

40. Towbin; Staehelin, T.; Gordon, J., et al. Electrophoretic Transfer of Proteins from Polyacrylamide Gels to Nitrocellulose Sheets: Procedure and Some Applications. *Proc. Natl. Acad. Sci. USA*. **1979,** *76*(9), 4350–4354.

41. Mullis, K. B. The Unusual Origin of the Polymerase Chain Reaction. *Sci. Am.* **1990,** *262,* 36–43.

42. Maxim, A. M.; Gilbert, W. A New Method for Sequencing DNA. *Proc. Natl. Acad. Sci. USA*. **1977,** *74*(2), 560–564.

43. Sanger, F. DNA Sequencing with Chain Termination Inhibitors. *Proc. Natl. Acad. Sci. USA*. **1977,** *74*(12). 5463–5467.

44. Nyrén, P. The History of Pyrosequencing. *Methods Mol. Biol.* **2007,** *373,* 1–14.

45. Shendure, J., Porreca, G. J.; Reppas, N. B.; Lin, X.; McCutcheon, J. P.; Rosenbaum, A. M.; Wang, M. D.; Zhang, K.; Mitra, R. D.; Church, G. M. Accurate Multiplex Polony Sequencing of an Evolved Bacterial Genome. *Science* **2005,** *309*(5741), 1728–1732.

46. Meyer, M.; Kircher, M. Illumina Sequencing Library Preparation for Highly Multiplexed Target Capture and Sequencing. *Cold Springs Harb Protoc.* **2010,** *2010*(6), pdb.prot5448.

47. Porreca, G. J. Genome Sequencing on Nanoballs. *Nat. Biotechnol.* **2010,** *28*(1), 43–44.

48. Eid, J.; Fehr, A.; Gray, J.; Luong, K.; Lyle, J.; Otto, G., et al. Real-time DNA Sequencing from Single Polymerase Molecules. *Science* **2009,** *323*(5910), 133–138.

49. Rusk, N. Torrents of Sequence. *Nat. Meth.* **2011,** *8*(1), 44.

50. Zheng, Z., et al. Titration-free Massively Parallel Pyrosequencing Using Trace Amounts of Starting Material. *Nucleic Acids Res.* **2010,** *38*(13), e137.

51. Shivashankar, H.; Nagaraj, R. B. G.; Ranganathan, S. A Hitchhiker's Guide to Expressed Sequence Tag (EST) Analysis. *Brief. Bioinform.* **2006,** *8*(1), 6–21.

52. McGuinness, M.; Dowling, D. Plant-associated Bacterial Degradation of Toxic Organic Compounds in Soil. *Int. J. Environ. Res. Public Health* **2009,** *6*(8), 2226–2247.

53. Harrison, M. M.; Jenkins, B. V.; O'Connor-Giles, K. M.; Wildonger, J. A CRISPR View of Development. *Genes Dev.* **2014,** *28*(17), 1859–1872.

54. Peng, Y.; Clark, K. J.; Campbell, A. M.; Panetta, M. R.; Guo, Y.; Ekker, S. C. Making Designer Mutants in Model Organisms. *Development* **2014,** *141,* 4042–4054.

55. Barrangou. R.; Doudna, J. A. Applications of CRISPR Technologies in Research and Beyond. *Nat. Biotechnol.* **2016,** *34,* 933–941.

56. Yang, L.; Güell, M.; Niu, D.; George, H.; Lesha, E.; Grishin, D., et al. Genome-wide Inactivation of Porcine Endogenous Retroviruses (PERVs). *Science* **2015,** *350*(6264), 1101–1104.

57. Larson, M. H.; Gilbert, L. A.; Wang, X.; Lim, W. A.; Weissman, J. S.; Qi, L. S. CRISPR Interference (CRISPRi) for Sequence-specific Control of Gene Expression. *Nat. Protoc.* **2013,** *8,* 2180–2196.

58. Nihongaki, Y.; Yamamoto, S.; Kawano, F.; Suzuki, H.; Sato, M. CRISPR-Cas9-based Photoactivatable Transcription System. *Chem. Biol.* **2015,** *22*(2), 169–174.

59. Welch, M.; Govindarajan, S.; Ness, J. E.; Villalobos, A.; Gurney, A.; Minshull, J.; Gustafsson, C. Design Parameters to Control Synthetic Gene Expression in *Escherichia coli*. *PLoS One* **2009a**, *4*, e7002.

60. Welch, M.; Villalobos, A.; Gustafsson, C.; Minshull, J. You're One in a Googol: Optimizing Genes for Protein Expression. *J. R. Soc. (Interface)* **2009b**, *6*(suppl. 4), S467–S476.

61. Costa, S.; Almeida, A.; Castro, A.; Domingues, L. Fusion Tags for Protein Solubility, Purification and Immunogenicity in *Escherichia coli*: The Novel Fh8 System. *Front. Microbiol.* **2014**, *5*, 63.

62. Young, C. L.; Britton, Z. T.; Robinson, A. S. Recombinant Protein Expression and Purification: A Comprehensive Review of Affinity Tags and Microbial Applications. *Biotechnol. J.* **2012**, *7*, 620–634.

63. Schunk, M. K; Macallum, G. E. Applications and Optimization of Immunization Procedures. *ILAR J.* **2005**, *46*(3), 241–257.

64. Frenzel, A.; Hust, M.; Schirrmann, T. Expression of Recombinant Antibodies. *Front. Immunol.* **2013**, *4*, 217. Published online: 2013 Jul 29. DOI:10.3389/fimmu.2013.00217.

65. Nascimento, I. P.; Leite, L. C. C. Recombinant Vaccines and the Development of New Vaccine Strategies. *Braz. J. Med. Biol. Res.* **2012**, *45*(12), 1102–1111.

66. Wang, J. W.; Roden, R. B. S. Virus-like Particles for the Prevention of Human Papillomavirus-associated Malignancies. *Expert Rev. Vaccines* **2013**, *12*(2), 10.1586/erv.12.151.

67. Curtiss, R 3rd, Xin, W.; Li, Y.; Kong, W.; Wanda, S. Y.; Gunn, B.; Wang, S. New Technologies in Using Recombinant Attenuated Salmonella Vaccine Vectors. *Crit. Rev. Immunol.* **2010**, *30*(3), 255–270.

68. Walsh, S. R.; Dolin, R. Vaccinia Viruses: Vaccines Against Smallpox and Vectors Against Infectious Diseases and Tumors. *Expert Rev. Vaccines* **2011**, *10*(8), 1221–1240.

69. Wells, J. M.; Mercenier, A. Mucosal Delivery of Therapeutic and Prophylactic Molecules Using Lactic Acid Bacteria. *Nat. Rev. Microbiol.* **2008**, *6*, 349–362.

70. Liljeqvist, S.; Ståhl, S. Production of Recombinant Subunit Vaccines: Protein Immunogens, Live Delivery Systems and Nucleic Acid Vaccines. *J. Biotechnol.* **1999**, *73*(1), 1–33.

71. Baeshen, N. A.; Baeshen, M. N.; Sheikh, A.; Bora, R.S.; Ahmed, M. M. M.; Ramadan, H. A. I.; Kulvinder, S. S., K. S.; Redwan, E. M. Cell Factories for Insulin Production. *Microb. Cell Fact.* **2014**, *13*, 141.

72. Lheureux, K.; Menrad, K. A Decade of European Field Trials with Genetically Modified Plants. *Environ. Biosafety Res.* **2004**, *3*, 99–107.

73. Boto, L. Horizontal Gene Transfer in Evolution: Facts and Challenges. *Proc. R. Soc. Biol. Sci.* **2010**, *277*(1683), 819–827.

74. Straughan, R. *Ethics, Morality and Animal Biotechnology*. BBSRC (Biotechnology and Biological Sciences Research Council), 1999. Available on: http://resources.schoolscience.co.uk/BBSRC/ethics/ethics_animal_biotech.pdf

75. Dorney, M. S.; Moore, V. The Regents of the University of California: Balancing the Need for Biotechnology Innovation Against the Right of Informed Consent. *Berkeley Tech. L. J.* **1990**, *5*(2), 334–368.

76. Liang, P.; Xu, Y.; Zhang, X.; Ding, C.; Huang, R.; Zhang, Z.; Lv, J.; Xie, X.; Chen, Y.; Li, Y; Sun, Y.; Bai, Y.; Songyang, Z.; Ma, W.; Zhou, C.; Huang, J. CRISPR/Cas9-mediated Gene Editing in Human Tripronuclear Zygotes. *Protein Cell* **2015,** *6*(5), 363–372.

77. Cyranoski, D.; Reardon, S. *Chinese Scientists Genetically Modify Human Embryos: Rumours of Germline Modification Prove True—and Look Set to Reignite an Ethical Debate.NatureNews,***2015**.doi:10.1038/nature.2015.17378.Availableon:https://www.nature.com/news/chinese-scientists-genetically-modify-human-embryos-1.17378.

CHAPTER 6

MOLECULAR DIAGNOSTICS

CONTENTS

ABSTRACT

The molecular basis of disease is increasingly being used for detection and diagnosis. The need of hour is early and accurate diagnosis of disease which can lead to adequate and accurate therapy. In addition, there is also a requirement of rapid and cost-effective technique which could be used on a daily basis. Last few years have been very fruitful in this context where research has led to establishment of such efficient technology which makes use of molecular basis of disease that is the biomarkers, such as PCR, Southern blot to name a few. An attempt has been made to include most of such techniques in this chapter for the benefit of reader.

6.1   MOLECULAR BASIS OF DISEASE

A disease can be defined as any abnormality in the living system. Causative agents for such kind of abnormality can be bacteria, virus, fungus, and parasites. Apart from this, mutation can also result in many types of disorders and/or disease manifestations. Sequencing of human genome has facilitated the identification of such sequences which have been altered or linked to a particular disease. Identification of the sequence is not enough for the mankind. To develop a therapy or cure, it is necessary to understand and identify the site of expression and its normal as well as altered function for a particular gene. Thus, it is very important to know the function of the gene in an unaltered cell so that the various mechanisms through which it works could be targeted to device cure or therapy. Molecular biology tools have been put to great use for all these purposes.

A timely diagnosis is the first and an important step in the prevention and treatment of these diseases. Although many laborious and costly clinical procedures were used initially, with the advancement of molecular biotechnology, various molecular diagnostic methods are now available and routinely used for the diagnosis and also treatment of these diseases. Among several methods used for the purpose, they have been selected and discussed in this chapter on the basis of their properties such as:

a)   They should be specific for the target molecule.
b)   They should have very high sensitivity and be able to detect even minute levels of the target.
c)   They should be easy to operate that is technical simplicity is required.

6.1.1 BIOMARKERS AND PERSONALIZED MEDICINE

Discussion about molecular diagnostics would be incomplete without mentioning biomarkers. These are signature molecules associated with many diseases such as cancer. These biomarkers have been put to great use for detection of various diseases and have paved the way for personalized medicine (PM) once diagnosis, prognosis, and selection of target therapies have been made. There are three broad categories of biomarkers as listed below:

a) General biomarkers
b) DNA biomarkers
c) DNA tumor biomarkers

It becomes pertinent to explain the role of biomarkers in PM, an emerging medical practice. PM makes use of an individual's genetic profile to make decisions regarding diagnosis, prevention, and treatment of diseases. PM is making a great advancement by making use of data from the Human Genome Project. Emphasis is laid on PM because not all therapies work for everyone. The various differences at the genetic level are responsible for a number of outcomes in treatment given for one disease to different persons. The availability of genomic data from the Human Genome Project is leading the researchers to find a precise and accurate treatment regime for many ailments customized for individuals. Thus, PM is beginning to achieve its goal of "the right therapy to the right patient at the right time." PM is helping us move closer to more customized, accurate, powerful, and predictable medication for an individual patient.

Therapies targeting biomarkers have been associated with several successes in the last decades, together with their limits and related unsolved questions. It has been suggested that most of the tumors will eventually develop resistance potentially due to intra-tumor heterogeneity and selection of additional molecular events for propagation and establishment of disease. As it is also known that in most of the tumors multiple rare genetic events occur, which proves to be the major hurdle to design an effective therapy. Administering one single medication will incur in nondurable results. So, the future lies in the fact that smart development in personalized therapies is the need of time which could change the regular history of several diseases, mainly tumors. It is too early to say that PM will change traditional therapy, however; in regards to biomarkers, it might create a safe and effective therapy for every individual patient. Knowledge of a patient's genetic profile can be of great help to the harried physicians for selection of proper medication or therapy and administer it using the proper dose or regimen. So before elaborating on the various tools of molecular diagnostics, it is necessary to know the utility of such genetic analysis and its various applications.

6.1.2   GENETIC ANALYSIS

Genetic testing is helpful in detecting if a person is carrying a specific altered gene (genetic mutation) which has been associated with a specific medical condition. Such test requires a sample of blood or tissue containing DNA. These tests are carried out for (a) diagnosis of a person with a genetic condition, (b) to help work out if the person has the chances of developing a disease, and finally (c) to determine if a person is a carrier of a certain mutation that is genetically inherited and could pass to their children. In some cases, genetic testing can be carried out to find out whether a baby is likely to be born with a certain genetic condition by testing samples of the fluid that surrounds the fetus in the womb (amniotic fluid) or cells that develop into the placenta (chorionic villi cells), which can be extracted from the mother's womb using a needle. Such analysis forms the basis of genetic counseling which is becoming a norm in today's time.

6.1.3   GENETIC COUNSELING

Genetic counseling is a service that provides support, information, and advice about genetic conditions. Genetic counseling takes into account various aspects as mentioned below:

- To learn about a particular health condition that runs in family, how it is inherited, and which family members are likely to be affected.
- Risk assessment of partners to pass on an inherited condition to the offspring is carried out.
- A family tree is constructed of both the partners after looking at the medical history.

- Support and advice is given to the family if having a child affected by an inherited condition and if they wish to have another child
- A discussion is held to educate about genetic tests, if appropriate, which include the risks, benefits, and limitations of genetic testing.
- This also helps in understanding the results of genetic tests and what they mean.
- Information is given about relevant patient support groups.

Preimplantation genetic diagnosis

For couples at risk of having a child with a serious genetic condition, preimplantation genetic diagnosis (PGD) may be an option. PGD involves using *in vitro* fertilization (IVF). IVF is a process in which eggs are removed from a woman's ovaries and fertilized with sperm in a laboratory. After a few days of fertilization, the resulting embryos are tested for a particular genetic mutation and a maximum of two unaffected embryos are transferred into the uterus. The advantage of PGD is to avoid the termination of fetuses affected by serious conditions. The major disadvantage of PGD lies in the fact that it has a modest success rate of achieving a pregnancy after IVF, and it involves substantial financial and emotional costs.

6.2   TOOLS OF MOLECULAR DIAGNOSTICS

Study of human genomics and genetics has become an important as well as an integral part of medicine and public health. For the patient diagnosis, determining the molecular basis of genetic disorders can be of great use. Apart from this, identification of familial mutations can lead to appropriate genetic counseling for families and possible prenatal diagnosis or PGD for future pregnancies.

Nucleic acid-based testing is becoming a crucial diagnostic tool not only in the setting of inherited genetic diseases (e.g., cystic fibrosis and hemochromatosis) but also in a number of neoplastic and infectious processes. Following diagnosis, molecular testing can help guide appropriate therapy by identifying specific therapeutic targets of several newly tailored drugs, thus playing an integral role in the application of pharmacogenomics (Fig. 6.1).

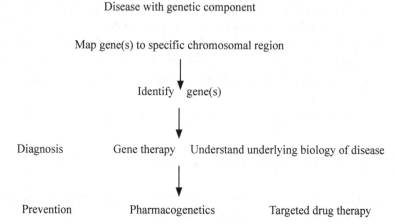

FIGURE 6.1   Steps involved in a genetic approach to the diagnosis and treatment of disease.

This tenet is fundamental to the newly emerging field of molecular diagnostics. Diagnosis, prognosis as well as the therapy can be based on the information obtained by making use of various molecular diagnostic tools capable of detecting any kind of pathogenic mutation in each sample. For analysis of patient samples in clinical diagnostic laboratories, many molecular biology techniques have been utilized. This chapter would deal with the development of molecular diagnostic approaches and some of the most commonly used assays as well as the most recent advances in this field.

The field of clinical molecular diagnostics has increased to a great deal in the last few decades, owing to advancements in human genetics basic research and technologies. In the early years, research laboratories primarily developed the techniques used to analyze genetic mutations. Many of these assays were then implemented into the clinical molecular diagnostic itinerary.

The initial assay techniques included diseases such as hemoglobinopathies and cystic fibrosis. These early molecular diagnostic methods detected indirect mutation through haplotype and linkage analyses which were labor consuming apart from requirement of large amounts of patient DNA. Another disadvantage was that extensive knowledge of the genomic region in question was of utmost importance, and the analysis did not always result in an easily interpretable result which proved to be a major setback in the widespread use of these early technologies.

Despite various drawbacks of the earlier technology, the foundation for molecular diagnostics as known and practiced today was laid by those very early beginning. Discovery of polymerase chain reaction (PCR) proved to be the game changer in molecular diagnostics. PCR technique was first described by Mullis et al. in 1986. PCR equipped the researchers as well as technicians to produce multiple copies of a DNA region, which eased the process of direct mutation detection/identification and led to faster analysis. Such was the impact of PCR technology that the assays being in use, prior to the discovery of PCR were adapted to use of PCR-amplified DNA instead of using genomic DNA which was a cumbersome procedure. PCR-based assays enabled us to analyze large number of samples in a very short time duration. Analysis of rare disorders along with the common ones was also accomplished by the use of PCR-based methods not possible with the earlier technologies. Today, with resources such as the 1000 Genomes Project, a detailed catalogue of human genetic variation, diagnostic molecular laboratories have access to the sequence of all human genes and a continuously growing database of human variation.

With the increasing popularity of next-generation sequencing (NGS) technologies, the most common technique still continues to be the automated Sanger sequence analysis for diagnosis of many genetic disorders clinically in molecular diagnostic laboratories. However, the assay choice often depends on the gene or alleles of interest and the number of patients to be screened. Generally, current molecular diagnostic assays are either targeted to specific alleles or analyze particular genes or groups of genes if there is no specific allele of interest. In this chapter, we describe some of the more common molecular techniques used in the analysis of both known and unknown mutations (Table 6.1) and discuss possible pitfalls of conventional PCR-based methodologies.

6.2.1   RESTRICTION FRAGMENT LENGTH POLYMORPHISM (RFLP)

This is among the pioneer techniques utilized in clinical molecular diagnostics for detection of genomic changes using Southern blotting and restriction fragment length polymorphism (RFLP). The Southern blot transfer hybridization assay was developed in 1975.[1] This was the time when complimentary DNA (cDNA) synthesis and cloning techniques

TABLE 6.1 Techniques Used in Clinical Diagnosis.

Techniques to detect known mutation	Techniques to detect unknown mutation	Techniques to detect gene copy number change
1. Southern/restriction fragment length polymorphism (RFLP)	1. Gradient gel electrophoresis (GGE)/ denaturing	1. Southern blot
2. Allele-specific oligonucleotide (ASO)	2. DGGE and temperature (TGGE)	2. Multiplex ligation-dependent probe amplification (MLPA)
3. Allele refractory mutation system (ARMS)	3. Single-strand conformation polymorphism (SSCP)	3. Array comparative genomic hybridization (aCGH)
4. Oligonucleoide ligation assay (OLA)	4. Heteroduplex analyses (HDA)	4. Single-nucleotide polymorphism (SNP) arrays
5. Pyrosequencing	5. Denaturing high-performance liquid chromatography (DHPLC)	
6. Real-time PCR	6. Protein truncation test (PTT)	
7. Sanger sequence analysis in case of known mutation	7. Sanger sequence analysis	

which provided the ability to determine the primary sequence of a number of genes[2] was also developed. All these developments led to identification of nucleotide sequence of many genes with the aid of cloned human cDNA. RFLP when used in conjunction with the previously described techniques along with the availability of the sequence of various genes provides a means to map mutant as well as normal genomic DNA. For example, genetic variations in a restriction enzyme site close to the beta-globin structural gene were identified only in people of African origin.[3] These polymorphic sites were put to use in the diagnosis of sickle cell anemia commonly found in the abovementioned ethnic group. These early studies helped in initiating the use of RFLP and Southern blotting in diagnostic tests such as linkage analysis and prenatal diagnosis. Disorders such as the thalassemia, cystic fibrosis, and phenylketonuria were among the first to be described.[4,5]

Flowchart to identify a disease-causing mutation in a gene using RFLP

↓

Cloning of gene of interest

↓

Development of probes for Southern blot analysis

↓

Digestion of genomic DNA with an array of restriction enzymes

↓

Detection of desired sequence polymorphic in size, with the help of DNA probes

↓

Identification of affected parents

↓

Polymorphic fragment size can be used for prenatal diagnosis

Apart from this, RFLP made it possible to analyze families that did not provide adequate information by single enzyme digestion about their haplotype and carrier status. If a particular abnormality is known to be associated with a particular haplotype, then this method could be applied to identify different forms of the disease as was done in case of β-thalassemia where different forms were present. Advantage of this technique is that it bypasses the need to repeatedly isolate the same mutation which was based on their haplotype background.[6] Later on, this procedure further gained importance with the advent of PCR amplification of DNA. RFLP of the PCR-amplified DNA region is now being used commonly, where mutation of interest is known and a restriction enzyme which cuts at the particular site is used (Fig. 6.2). Patients that were carriers are distinguishable from those that were either homozygous wild type or homozygous mutant by the banding pattern of the PCR products on a gel. For PCR-based RFLP analysis, some of the first applications were in the characterization of sickle cell anemia alleles.[7]

FIGURE 6.2   Agarose gel picture showing marker in the first lane and restriction enzyme-treated PCR product in the subsequent lanes.

6.2.2 ALLELE-SPECIFIC OLIGONUCLEOTIDE HYBRIDIZATION (ASO) FOR DETECTION OF SPECIFIC MUTATION ASSOCIATED WITH A DISORDER

This assay is an advanced technique in the sense that it is able to detect specific mutations in a specific disorder. ASO hybridization, commonly called dot blot analysis is based on the principle that even a single-base-pair change between a target region and the probe is capable of destabilizing the hybrid when probing a particular region of DNA.

Flowchart: Development of probes (dual probes one complementary to the wild-type allele and the other to the mutant allele).

Digestion of DNA and separation (gel electrophoresis)
↓
Immobilization of digested DNA on a membrane
↓
Hybridization with radioactively labeled probe

Analysis of hybridization is based on the fact the heterozygous condition is represented by reaction with both the probes. Reaction with only one probe signifies homozygous condition for either wild type or mutant allele.

Initially, the probes developed were radiolabeled which later on gave way to the development of detection techniques using colorimetry. This was accomplished by conjugating the probes with biotin which facilitated the detection using streptavidin conjugated to horseradish peroxidase, thus eliminating the need for radioactivity and making use of colorimetric or chemiluminescent detection. This technique was used in the early 1980s in the detection of allele for the sickle cell anemia[8] and subsequently prenatal diagnosis of β-thalassemia.[9] ASO was also a technique that benefitted from the use of PCR. Instead of probing nonamplified genomic or cloned DNA, regions of interest were PCR amplified first and then probed, thus making the process faster and more efficient. As the addition of PCR to ASO facilitated a more rapid detection of mutations[10] and became one of the most widely used and sort after techniques to study targeted alleles. Screening of large number of samples for few mutations made use of this technique which was an added advantage. For this, each oligonucleotide probe was labeled separately. Modification also allows for simultaneous assay of multiple mutations in a variety of genes in a single patient sample thus reducing the time for detection to multifold.

6.2.3 AMPLIFICATION REFRACTORY MUTATION SYSTEM TO DETECT KNOWN POINT MUTATION

Amplification refractory mutation system (ARMS) is also a PCR-based method and could be used to analyze known point mutations. ARMS is based on the fact that DNA amplification is inefficient due to existence of mismatch between the template DNA and the 3' terminal nucleotide of a PCR primer.[11] For example, a primer with a 3' terminal nucleotide that is complementary to the wild-type allele will not have efficient extension when a mutation is present (causing a mismatch) and vice versa. Thus, differentiation between two alleles can be done by simple PCR amplification. A primary function of the alleles of interest is the design and optimization of ARMS assays and the nucleotides surrounding them. Often, incorporating additional mismatched nucleotides near the target allele can enhance the reaction.[11] Many mutations can be analyzed by using multiple sets of primer pairs which allow simultaneous analysis of many mutations at a time. This technique has been used to identify patients who carry known mutations in many disorders which include cystic fibrosis and phenylketonuria.[12]

6.2.4   OLIGONUCLEOTIDE LIGATION ASSAY: AN ADVANCEMENT USING COMBINATION OF PCR AND LIGATION

The oligonucleotide ligation assay (OLA) combines PCR with ligation at a target allele site in one reaction. To the PCR amplified target region, three oligonucleotides are added via enzymatic ligation. Three probes are used for this assay, first one called the reporter is a common probe used and that is complementary to the target DNA sequence which is immediately 3'to the allele of interest. The other two known as the "capture" probes used are complementary to the target DNA sequence immediately 5'to the target allele. The "capture" probes differ only in their final 3'terminal nucleotide of the target allele. Only if there is a perfect match between the capture probe and the target allele, ligation between the reporter and the capture probes can occur thus enhancing the efficiency of the process. Several different detection methods for OLA have been developed including detecting different lengths of ligated products for the target alleles, as well as alternate labels on the capture probes such as fluorescence or biotin[13] (Fig. 6.3).

OLA has an advantage of being rapid and sensitive detection technique with a significantly reduced cost. OLA has been utilized in the detection of mutations in number of metabolic disorders, cystic fibrosis, and pharmacogenetics to name a few.[14–16]

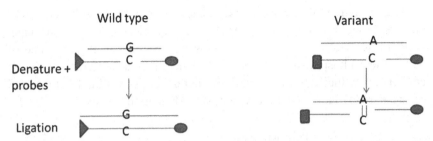

Wild type   Variant

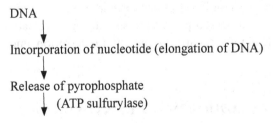

FIGURE 6.3 A pair of allele-specific probes specific for wild-type allele and another specific for variant allele along with a fluorescent common probe is taken. The 3' ends of the allele-specific probes are immediately adjacent to the 5' end of the common probe. In the presence of thermally stable DNA ligase, ligation of the fluorescence-labeled probe to the allele-specific probe occurs only when there is a perfect match between the variant or the wild-type probe and the PCR product template. These ligation products are then separated by electrophoresis which permits the recognition of the wild-type genotypes, the variants, the heterozygous, and the unligated probes.

6.2.5 PYROSEQUENCING: THE USE OF LUMINESCENCE FOR REAL-TIME DETECTION

This technique was first described in 1985 and is said to be an enzymatic method used for monitoring DNA polymerase activity.[17] Pyrosequencing is a DNA sequencing technology which is based on real-time detection of DNA synthesis that can be monitored by luminescence (Fig. 6.4).

Flowchart of the process of pyrosequencing: Conversion of luciferin to oxyluciferin by ATP formed by the action of ATP sulfurylase

DNA
↓
Incorporation of nucleotide (elongation of DNA)
↓
Release of pyrophosphate
↓ (ATP sulfurylase)
Conversion of pyrophosphate to ATP in the presence of adenosine 5' phosphosulphate
↓← ATP
Conversion of luciferin to oxyluciferin (generation of visible light which can be measured) by ATP

The principle of the assay is a reaction in which each sequential nucleotide incorporated during DNA synthesis releases a pyrophosphate. ATP sulfurylase converts that pyrophosphate to ATP in the presence of adenosine 5'phosphosulfate. That ATP then drives a luciferase-mediated conversion of luciferin to oxyluciferin that generates visible light that can be measured (Fig. 6.4). Unincorporated dNTPs are degraded.

$$NA)n + Nucleotide \xrightarrow{\text{Polymerase}} (NA)_{n+1} + PPi$$

$$PPi + APS\ ATP \xrightarrow{\text{sulfurylase}} ATP + SO_4^{2-}$$

$$ATP + Luciferin + O2 \xrightarrow{\text{Luciferase}} AMP + PPi + Oxyluciferin + CO_2 + Light$$

FIGURE 6.4 Depiction of the general principal of pyrosequencing reaction system. A polymerase catalyzes incorporation of nucleotides into a nucleic acid chain resulting in release of a pyrophosphate (PPi) molecule and its subsequent conversion to ATP which is catalyzed by ATP sulfurylase. Luciferase reaction aided by the ATP formed results in the production of measurable light during which a luciferin molecule is oxidized.

The nucleotides are added in a specific order such that the pattern for the wild type or mutant allele is expected. Advantage of pyrosequencing over other sequencing techniques such as that described by Sanger earlier is that it facilitates short read length when analysis of genetic variants for applications such as single-nucleotide polymorphism (SNP) genotyping or detection of known mutations is required. Pyrosequencing is greatly used in an array of tests including pharmacogenetic testing in the analysis of polymorphisms particularly in genes involved in drug metabolism.[18,19] Such analysis of polymorphism provides information on metabolism of any drug by a person/patient thus enabling clinicians to make more informed decisions regarding patient prescription and dosage.

6.2.6 REAL-TIME PCR: AN ADDITION TO CONVENTIONAL PCR

In the mid-1990s, a technique involving the analysis and quantification of DNA or RNA in real time was developed (Fig. 6.5).[20,21] This is a very sensitive assay which enables us to make an accurate quantification of a PCR product during the exponential phase of PCR. The first real-time PCR was performed using TaqMan probes.[20,21] These probes specifically hybridize to the region around a target allele internal to the primer-binding sites. The TaqMan probe is generally ragged at each end with a fluorescent molecule, a reporter dye, and a quencher. The quencher prevents the reporter from fluorescing as long as the two are in close proximity. As the cycle of PCR progresses, the probe is degraded by the exonuclease activity of Taq polymerase, separating the fluorophore from the quencher and allowing fluorescence emission. This emission can then be measured. The increase in fluorescence at every cycle is measured and correlates directly to the amount of PCR product formed.[21,22]

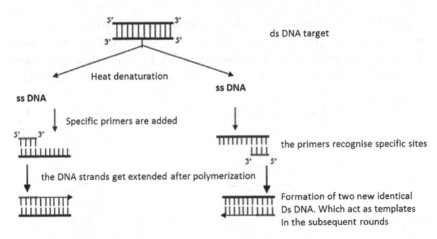

FIGURE 6.5   Basic steps of a PCR/reverse transcriptase PCR reaction.

Another type of real-time PCR uses fluorescent DNA-intercalating dyes. The first use of this method measured the increase in ethidium bromide fluorescence in double-stranded DNA molecules and was referred to as kinetic PCR.[20]

Later, SYBR Green I was used because compared to ethidium bromide, it is less toxic and incorporates into double-stranded DNA. During the PCR reaction, the amount of double-stranded DNA increases exponentially with the corresponding increase in the amount of dye incorporation and emission, which can be a measure. This technique requires very little DNA manipulation, besides its ability to measure factors such as DNA copy number in real time rapidly makes this assay preferable in molecular diagnostic laboratories.

6.2.7   GRADIENT GEL ELECTROPHORESIS

Gradient gel electrophoresis (GGE) which includes temperature gradient gel electrophoresis (TGGE) and denaturing gradient gel electrophoresis (DGGE) is based on the principle that the electrophoretic mobility of double-stranded DNA fragments is affected and altered by their partial denaturation state (Fig. 6.6). The technique was first used in characterizing human mutations to detect mutation causing β-thalassemia.[23] GGE made possible the detection of allelic changes even without the knowledge of the exact DNA sequence of the region in question and simultaneous screening of multiple nucleotide changes in a single region which could not be done in other techniques such as ASO. In GGE, PCR amplified-fragments are denatured, reannealed, and then analyzed on denaturing gradient gels in this technique by observing their mobility. The mobility of the fragments through the gel is based on their melting temperatures (T_m).

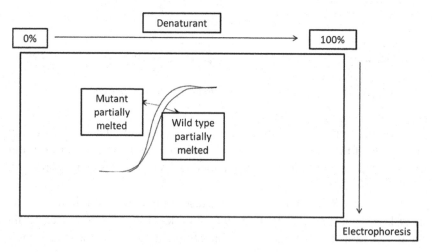

FIGURE 6.6   Denaturating gradient gel electrophoresis.

The dissociation and mobility can be altered even with a single-nucleotide substitution because the T_m is dependent on overall DNA sequence. Heteroduplexes of wild-type and mutant DNA fragments generally migrate slower than homoduplexes in polyacrylamide gels under denaturing condition due to mismatching of alleles and can therefore be separated by gradients of linearly increasing denaturant such as urea (DGGE) or temperature (TGGE). Both TGGE and DGGE require a gradient of either temperature or denaturant for detection.

Yoshino et al. introduced for the first time the temporal temperature gradient gel electrophoresis (TTGE) as a modification of TGGE.[24] TTGE allows for easier temperature modulation because the temperature of a gel plate increases gradually and uniformly with time. This increases the sensitivity as the separation range expands. One of the first reports showing successful application of this method to clinical diagnosis was in the detection of mutations in mitochondrial DNA.[25] Subsequently, TTGE has been used as a method of detection for germline mutations in a variety of disorders including cystic fibrosis[26] and somatic mutations in cancer tissues.[27]

6.2.8   SINGLE-STRAND CONFORMATION POLYMORPHISM AND HETERODUPLEX ANALYSES

Single-strand conformation polymorphism (SSCP) and heteroduplex analyses (HDA) were developed soon after the introduction of PCR technology.[28,29] HDA is based on PCR product migration through a nondenaturing gel in a fashion similar to GGE in which heteroduplexes are analyzed in relation to homoduplexes. These heteroduplexes are formed by mixing denatured, single-stranded mutant and wild-type DNA PCR products, followed by reannealing them to form duplexes room temperature slowly. The differential migration of these duplexes depends on whether they are heteroduplexes or homoduplexes of wild-type or mutant PCR fragments. Therefore, mutations can quickly be detected through simple gel migration analysis (Fig. 6.7). During electrophoresis, single-stranded DNA fragments form specific three-dimensional shapes and exhibit a unique conformation according to their nucleotide sequence. Therefore, their electrophoretic mobility is dependent upon their three-dimensional shape (Fig. 6.7). Thus, even a single base difference between a DNA

fragment being tested and its wild-type counterpart is sufficient to have a
different conformation and as a result migrate at a different position during
electrophoresis (Fig. 6.7). An advancement of this was Fluorescent SSCP
(F-SSCP) which used fluorescently labeled PCR products and an automated DNA sequencer which was developed in the early 1990s[30] further
increased the sensitivity and efficiency of this technique.

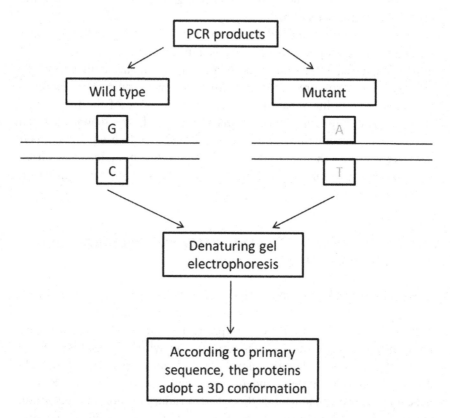

FIGURE 6.7 PCR amplicons of wild-type allele and mutant allele are denatured and
electrophoresed on a nondenaturing polyacrylamide gel.

One of the first reported uses of HDA in molecular diagnostics was
in the detection of three base pair deletion in cystic fibrosis.[31] These
techniques have been used for years in the detection of mutations in a
number of disorders including a variety of cancers, phenylketonuria, and
retinoblastoma.[32–34]

6.2.9   DENATURING HIGH-PERFORMANCE LIQUID CHROMATOGRAPHY

Denaturing high-performance liquid chromatography (DHPLC) is a sensitive technique having advantage over many techniques and was first reported in 1997.[35]

i)   DGGE is labor intensive and requires process optimization and analysis by gel electrophoresis, so DHPLC is a better option

ii) SSCP and HDA lacked sensitivity, thus in this case too DHPLC is a better technique.

Briefly, like HDA, DNA homo- or heteroduplexes are formed after they are denatured and allowed to reanneal. DNA duplexes are subjected to a positively charged chromatography column. The binding of fragments to the column occurs at different strengths depending whether they are homo- or heteroduplexes and generate distinct chromatographic patterns as they elute from the column at different times and thus could be analyzed.

6.2.10   THE PROTEIN TRUNCATION TEST

The protein truncation test (PTT), also known as the *in vitro* synthesized protein assay (IVSP) is based on the change in protein sizes which is a result of *in vitro* transcription and translation of a gene.[36] The PTT was initially developed in the early 1990s to detect early termination mutations in the dystrophin gene responsible for Duchenne and Becker muscular dystrophy.[37] PTT is most frequently applied for the detection of premature truncation mutations in cancer-causing genes in which many truncating mutations have been identified.[38,39]

In this technique, RNA template is subjected to reverse transcription which generates a cDNA copy. cDNA is amplified with the aid of primers which have been designed for *in vitro* transcription and translation. The resulting proteins are then analyzed by SDS-PAGE electrophoresis. Translation products derived from truncating frameshift or nonsense mutations result in proteins of mass lower than the expected full-length protein.

6.2.11   SANGER SEQUENCING

Although all the mutation scanning methods described in the above sections are relatively easy to perform and fairly sensitive, they often require an extensive amount of design and optimization. In addition, any mutations detected by these scanning methods need to be ultimately confirmed by Sanger sequencing. Therefore, nowadays, the capillary electrophoresis-based Sanger sequencing has become the most widely used approach for DNA analysis in molecular diagnostic laboratories. Sequencing was first described by Sanger et al. in 1977.[40] The clinical assays based on Sanger sequencing use PCR-amplified products of a particular region of interest. If a specific target allele is of interest, a single PCR product or all coding exons plus flanking intronic sequences of a gene may be used. Each amplicon must then be sequenced independently. For clinical diagnosis in labs, 2x coverage of a sequence is generally required. Often this is accomplished by sequencing once each in the forward and reverse directions. However, sometimes this is not possible due to repetitive sequences around the region of analysis; for this, two separate forward or reverse sequences are used. Therefore, although it is faster, safer, and often cheaper than some of the other techniques, it is still somewhat laborious and incurs a high cost of operation. In recent years, "next-generation" or massively parallel sequencing technologies have been developed that will provide clinical diagnostic laboratories the ability to offer analysis of multiple genes or even the whole exome at a cost that is competitive with single-gene Sanger sequence analysis.

Detection of Copy Number Variations

Chromosome analysis is important in the diagnosis of conditions such as congenital abnormalities, intellectual disability, and developmental delay. Routine chromosomal analysis can detect both balanced and unbalanced structural rearrangements as well as deletions and duplications larger than ~5 Mb in size, in addition to whole chromosome aneuploidy. However, conventional karyotype analysis is unable to detect sub-microscopic deletions and duplications that are a common cause of intellectual disability. While Southern blotting is able to detect copy number changes in a number of genes, as previously discussed, it is very labor intensive, requires large amounts of DNA, and is only able to analyze a single region at a time. In addition, Southern blot may not be able to detect copy number changes in small regions such as single exon deletions. For this, real-time PCR can be used to detect copy number changes in small regions, although its use in a multiplex assay is also limited by the number of fluorescent dyes available and quantification of the data can be problematic if multiple primer pairs are desired. Therefore, a variety of techniques have been developed to increase the resolution of detection for chromosomal alterations, such as multiplex ligation-dependent probe amplification (MLPA), array comparative genomic hybridization (aCGH), and SNP arrays.

6.2.12 SOUTHERN BLOTTING

Southern blotting and RFLP are commonly used to track mutations in various disorders. Although RFLP analysis became PCR based later on, Southern blotting techniques still provides additional information for certain diseases. The fragile X mental retardation syndrome was one of the earliest disorders which were clinically diagnosed through Southern blotting and RFLP techniques. Mapped in the late 1980s and early 1990s using linkage and RFLP [41-45], analysis of the number of CGG repeat expansions and their methylation status in the 5' untranslated region of the FMR1 gene is one of the most common assays performed in clinical diagnostic laboratories today. PCR-based methods can be used to amplify the region containing the repeats, which can be indicated by the size of the PCR product. The efficiency of the reaction is somewhat inversely related to the number of repeats and the larger the size, the more difficult it is to PCR amplify the repeat. In addition, PCR does not provide any methylation information. Southern blotting allows the determination of both the repeat region size and its methylation status at the same time (Fig. 6.8). During restriction enzyme digestion, methylation-sensitive restriction enzymes can be used to distinguish between methylated and unmethylated species. Even though it is laborious and requires a large amount of DNA, the Southern blot is still used today in many clinical molecular diagnostic laboratories in the analysis of many diseases in particular trinucleotide repeat expansion disorders.

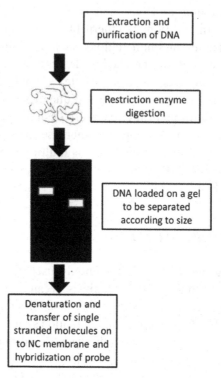

FIGURE 6.8 Southern blot technique for detection of mutation in DNA sample.

6.2.13 MULTIPLEX LIGATION-DEPENDENT PROBE AMPLIFICATION (MLPA)

Many diseases are an outcome of change in the copy number of various sequences. Diagnosis of such diseases requires techniques which can pin point any change in the copy number. For many clinical diagnostic laboratories, MLPA is an assay used for the detection of copy number variations. MLPA has the advantage of analyzing multiple regions of interest simultaneously with a low operating cost. Briefly, MLPA is essentially a combination of two techniques:

   a) Amplified fragment length polymorphism (AFLP) in which up to 50 different multiple DNA fragments are amplified in a single reaction with a lone primer pair.

b) Multiplex amplifiable probe hybridization (MAPH) in which multiple target oligonucleotide probes are hybridized to specific nucleotide sequences [46]. These probes are then also amplified with a single primer pair.

MAPH, like the Southern blotting technique, also requires the immobilization of samples to a membrane and removal of unbound probes through multiple washing steps. The benefit of using MLPA is that it allows for the amplification of multiple oligonucleotide probes in a single reaction without the immobilization of sample to a membrane, besides, excess probe removal is not an absolute requirement. In the MLPA technique, each probe set consists of two oligonucleotides capable of hybridizing to adjacent sides of the target sequence. Only when both oligonucleotides bind to the correct nucleotide sequence can they be ligated into a single probe. Each probe set is highly specific for detection and gives rise to a unique amplification product of a particular size that can then be separated by capillary electrophoresis. Cost effectiveness and fast result are other major advantages of using MLPA in a clinical laboratory setting as the universal tags used allow multiple amplicons to be produced in a single reaction.

However, MLPA does have its disadvantages which are

a) Due to the limits of multiplexing, each gene in a kit generally only has a limited number of probes per exon, with maximum probe number of about 50 per reaction.
b) Single nucleotide polymorphisms (SNPs) in probe regions can cause a decrease in the binding efficiency of oligonucleotides and give false positive results.
c) Further studies requiring additional PCR reactions are necessary in situations when a deletion is detected and the break points of that deletion are desired.
d) Extensive design "rules" sometimes make the development of an MLPA assay difficult.

6.2.14   ACGH TECHNIQUE

This technique was described in the early 1990s [47], and is now widely being used for the identification and characterization of chromosomal abnormalities in many different cell types. CGH analysis works on the principle of detection of chromosomal deletions and duplications by comparison of equal amounts of genomic DNA from a patient and a normal control.

Briefly, patient and control DNA are each labeled with a different fluorescent dye. Equal amounts of labeled patient and control DNA are then mixed together and co-hybridized to the array, which is a microscope slide onto which small DNA fragments (targets) of known chromosomal location have been affixed. In a particular region, if the oligo density is high enough, even small single exon deletions and duplications may be detected.[48,49] Some of the first clinical diagnostic arrays were constructed using bacterial artificial chromosome (BAC) clones as targets.[50]

6.2.15 SINGLE-NUCLEOTIDE POLYMORPHISM ARRAYS

SNP arrays were originally designed to genotype human DNA by simultaneously analyzing thousands of SNPs across the genome.[52] Since their inception, SNP arrays have been used for several other applications including absence of heterozygosity and detection of copy number changes. SNP arrays, like a CGH, are also based on oligonucleotide probes immobilized to glass slides. However, SNP arrays use only a single patient DNA, unlike CGH arrays that use both patient and control samples for comparison. There is a differential binding of the patient DNA to the oligonucleotide probes depending on the target SNP allele. Therefore, the resolution of the array is limited by SNP distribution. One major advantage of SNP arrays is their ability to detect copy number neutral differences in cases of absence of heterozygosity (AOH) that may occur as a result of uniparental isodisomy (UPD) or consanguinity (two copies), or deletion (one copy) such as loss of heterozygosity (LOH) associated with tumors. They can also detect copy number variants, but do not have the exon-by-exon coverage that most CGH arrays have nowadays.

Conventional PCR-Based Methods: Disadvantages

When a single set of primer is used in for clinical molecular diagnostic laboratories, and if an SNP present within a primer site it may disrupt the binding of that primer, and allele dropout could unknowingly occur. This can result in misinterpretation of the result obtained. Secondly if the region of interest contains a mutation, it may be missed during amplification. For example, if a heterozygous deletion encompasses the region of amplification, only one chromosome will be amplified which would also result in incorrect analysis. Failure of allele amplification on one chromosome will cause heterozygous mutations also to appear homozygous. These problems can be minimized by continuous reassessment of the presence of SNPs in primer sites using the constantly updated SNP database. In addition, whenever possible, parental testing must be carried out if a homozygous mutation is identified in an apparently homozygous point mutation in an affected proband with an autosomal recessive disease. If testing of the parents does not confirm their carrier status, additional molecular analyses can be performed to identify the underlying molecular etiology.[53] For any PCR-based analyses, firstly, allele dropout due to SNPs at primer sites should always be ruled out. Capture-based NGS do not rely on PCR primers, thus, will not have the problem of allele dropout. However, due to high GC content or the interference of pseudogenes, some regions of the genome may have poor coverage which will require Sanger sequence analysis. All positive results obtained by NGS should be confirmed by a secondary method, which is usually Sanger sequencing.

6.2.16 MICROARRAY ANALYSIS (DNA CHIP TECHNOLOGY)

This technique facilitates the analysis of thousands of mRNA in a parallel fashion, unlike the abovementioned techniques because the other techniques provide expression measurements on defined sets of genes. Chip techniques generally involve the arraying and deposition of target cDNAs or oligonucleotides onto support matrices. These matrices include nylon filters, microscope slides, or silicon chips. Differential mRNA analysis using DNA arrays is done by hybridizing the target array against total cDNA pools which has been derived from particular cell or tissue types of interest (as in case of diseased tissue). Radiolabeled tracers or fluorescent tags are used to quantitate the quantity of probe bound to each target cDNA on the array which is proportional to the quantity of corresponding RNA in the cells/tissue from which it was derived. The overall hybridization to the array provides a profile of the relative message levels for each gene represented on the microarray (Fig. 6.9). Differential analysis is achieved by comparing the hybridization signals generated by each cDNA/mRNA sample at each spot on the array. For instance, signal derived from diseased versus normal tissues. Thus, a large portion of the expressed genome can be compared in a single experiment because high-density microarrays can currently contain up to 10,000 distinct genes. This technology also helps investigators to understand the pharmacological and toxicological impact of new drugs on human tissues, identify therapeutic targets, and develop new approaches for diagnosing disease.

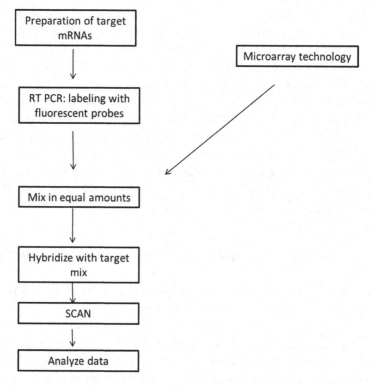

FIGURE 6.9 Schematic representation of microarray technique. mRNA from more than one sources is tagged with fluorescent tags with the aid of reverse transcription and mixed in equimolar concentration. This mixture is hybridized to a slide which has been pre-spotted with specific cDNA. Hybridization is followed by scanning and data analysis. The pattern of hybridization is characteristic activity signature of the subject and ratio of two signals can be calculated for determination of gene expression.

6.2.17   FLUORESCENT IN SITU HYBRIDIZATION (FISH)

FISH is based on the use of fluorescence-labeled oligonucleotide probes that specifically attach to their complementary DNA sequence target on the genome and label that region with fluorescence color (e.g., acridine orange, texas red, FITCI green). The labeled region can then be easily visualized under a fluorescence microscope (Fig. 6.10). Currently, three types of probes are in wide use:

- Painting probes that identify an entire chromosome by attaching to overlapping sequences on its target (e.g., chromosome 17) and thus "painting" that chromosome with the chosen fluorescence color.
- Centromeric probes that identify a specific chromosome's centromeric region of and thus help enumerate the copy number of that chromosome even in a nondividing cell (interphase state).
- Allele-specific probes that adhere to a specific target allele sequence such as the tumor suppressor gene or the oncogene.

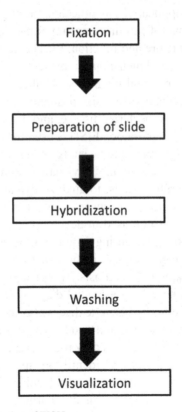

FIGURE 6.10   A layout plan of FISH.

FISH has several advantages over many other conventional cytogenetics methods in "gene malfunctioning" studies due to amplification, chromosomal deletions, and translocations. Secondly, conventional cytogenetics is a time-consuming cell culture step which needs to be performed mainly with fresh tissue samples, whereas FISH, on the other hand, can be performed on cells in dividing (metaphase) as well as resting (interphase) stages. It can be performed on fresh frozen as well as archival cytologic smears or paraffin-embedded tissue sections. This great versatility, in addition to the topographic advantage of fluorescent microscopic examination, which allows distinction between signals from tumorous and nontumorous cells, has fuelled the field of "interphase cytogenetics" in both tumor and prenatal settings.

FISH is often used in evaluation of breast carcinoma *HER2/neu* onco-gene amplification and for different translocation detections in acute and chronic myelogenous leukemia.

FISH has several applications in molecular biology and medical science, which include diagnosis of chromosomal abnormalities, gene mapping, and studies of cellular structure and function. FISH can be used for "painting" Chromosomes in three dimensionally preserved nuclei; for diagnosis of infectious disease, prenatal diagnosis of inherited chromosomal aberrations, tumor cytogenetic diagnosis, postnatal diagnosis of carriers of genetic disease, viral and bacterial disease, and detection of aberrant gene expression clinically. In laboratory, FISH is used for various purposes such as mapping of chromosomal genes, study of the evolution of genomes, to analyze nuclear organization, to visualize chromosomal territories and chromatin in interphase cells, to analyze dynamic nuclear processes, somatic hybrid cells, replication, chromosome sorting, and lastly to study tumor biology. It can also be used in developmental biology to study the temporal expression of genes during differentiation and development.

Recent revolutionary progress in human genomics is has reframed our approach to therapy and diagnosis. Tests which make use of Nucleic acid-based testing is becoming a crucial diagnostic tool day by day, not only in the setting of inherited genetic disease (as in case of cystic fibrosis and hemochromatosis) but also in a wide variety of neoplastic and infectious processes. Diagnosis, followed by molecular testing helps greatly in guiding for appropriate therapy by identifying specific therapeutic targets of several newly tailored drugs. Thus, molecular diagnostics is playing an important role in the application of pharmacogenomics. Molecular diagnostics provides the necessary background for any successful application of genetic therapy or modifiers of biologic response. It is a great tool for prognosis of disease and response toward therapy and detection of minimal residual disease.

6.3   GENE THERAPY

Since a number of disease have been shown to have molecular basis the next logical work would be to aim for gene therapy, which in a layman term means to rectify the altered gene. This needs sequencing and cloning of the normal gene to treat a genetic disease. The defect in individuals who have a mutant form of that gene can be corrected by this cloned normal gene. For this, a normal functioning gene is added to the defective cells, thereby providing the required protein to correct the genetic disease. In addition, in some diseases, it will be necessary to prevent the over expression of a deregulated normal gene. The three approaches followed for gene therapy are as follows:

1)   *Ex vivo* gene therapy
2)   *In vivo* gene therapy
3)   Antisense therapy

6.3.1   EX VIVO GENE THERAPY

Somatic cells from an affected individual are collected. The isolated cells are grown in culture. These cells are then transfected by retroviral cloning vectors containing the remedial gene construct. The cells are further grown and those cells which contain the gene of interest are selected and finally transplanted or transfused back into the patient. These transplanted transfected cells will synthesize the gene product, that is, the protein. For example, sickle cell anemia, thalassemia, etc. have been cured by this type of treatment.

6.3.2   IN VIVO GENE THERAPY

In this type of treatment, there is the direct delivery of the remedial gene into the cells of a particular tissue of the patient, using retroviral vectors, or even plasmid DNA constructs. This type of treatment is used to cure neuronal degeneration, muscular dystrophy, and brain cancer patients.

6.3.3 ANTISENSE THERAPY

Antisense therapy is aimed to prevent or lower the expression of a specific gene (Fig. 6.11). Some type of genetic diseases and cancers are associated with dysregulation or over expression of the genes which results in production of gene product in an excess amount or its continuous presence in the cell leading to disruption of the normal functioning of the cell. In such diseases, normal gene addition will not be sufficient, instead it would be more useful to block the synthesis of the gene product (protein). Thus, in antisense therapy a nucleic acid sequence complementary to complete or a part of that specific mRNA is introduced into the target cell. Hence, the mRNA produced by the normal transcription of the gene will hybridize with the antisense oligonucleotide by base pairing, thereby preventing the translation of this mRNA, resulting in reduced amount of target protein. The antisense therapy is used in treatment of sickle cell anemia, various cancers, atherosclerosis, AIDS, and leukemia.

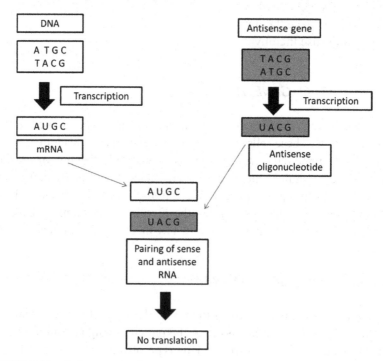

FIGURE 6.11   Antisense gene therapy.

6.4   LEGAL AND ETHICAL ISSUES ASSOCIATED WITH DNA TESTING

There are many issues related to DNA testing which needs to be addressed, such as medicine, public health and various policy regarding the specifications of circumstances under which these tests could be undertaken. The various uses of the result obtained from such testing needs to be defined. Another serious question is, should people be allowed to choose or refuse the test or should it be made compulsory for all for example in a newborn screening. Should the people be able to control the access to the results of the test lest the third party such as employers or insurers take advantage of this information. It is very important that people are not treated unfairly because of their genotype.

The resolution to the abovementioned problems can be achieved by the important ethical and legal principles which are

a) Autonomy implies that the person has the right to control the future use of genetic material which he/she might have submitted for particular analysis besides the clause when the genetic material itself and the information derived from that material may be stored for future analysis, such as DNA bank.

b) Confidentiality, this term implies that access to sensitive information must be controlled and limited to parties authorized to have the access. Information provided is given in confidence that it would not be disclosed to others or if at all disclosed then would be only within certain limits. This state of nondisclosure or limited disclosure may be protected by social, moral, legal principles and rule.

c) Privacy can be defined as a state of limited access to a person. People can be said to have privacy if others do not have access to them. This is very important to safeguard the interest of a person from anauthorized intrusion by others. If a person undergoes genetic tests it provides the power to make an informed and independent decision about whether and which others can know the details of the test performed and result. The others include insurers, employers, spouses, other family members, social agencies, researchers, educational institutions, etc. to name a few.

d) Equity—Certain modification in legislatures have been made to prohibit any discrimination based on genotype, for example, in employment. There is a provision to give needy people health care which can include some genetic services under government program. Many countries provide free health cars to citizens above 65 years of age.

6.5   REVIEW QUESTIONS

1) What is molecular diagnostics?
2) How molecular diagnostic approach has an edge over conventional diagnostic techniques?
3) What is PCR and how it is applied in various molecular diagnostic tools?
4) What do you understand by single nucleotide polymorphism and its application in molecular diagnostics?
5) What are the methods used when sequence is not known?
6) What is pyrosequencing?
7) What is the technique used when thousands of sequence have to be analyzed in a parallel fashion, and explain the principle of this tool.

KEYWORDS

- **molecular biology techniques**
- **biomarkers**
- **genetic profile**
- **molecular diagnostics**
- **cloning vectors**
- **PCR**

REFERENCES

1. Southern, E. M. Detection of Specific Sequences among DNA Fragments Separated by Gel Electrophoresis. *J. Mol. Biol.* **1975,** *98*(3), 503–517.

2. Rougeon, F.; Mach, B. Stepwise Biosynthesis *in vitro* of Globin Genes from Globin mRNA by DNA Polymerase of Avian Myeloblastosis Virus. *Proc. Natl. Acad. Sci. USA* **1976,** *73*(10), 3418–3422.

3. Kan, Y. W., Dozy, A. M. Polymorphism of DNA Sequence Adjacent to Human Beta-globin Structural Gene: Relationship to Sickle Mutation. *Proc. Natl. Acad. Sci. USA* **1978,** *75*(11), 5631–5635.

4. Woo, S. L.; Lidsky, A. S.; Guttler, F.; Chandra, T.; Robson. K. J. Cloned Human Phenylalanine Hydroxylase Gene Allows Prenatal Diagnosis and Carrier Detection of Classical Phenylketonuria. *Nature* **1983,** *306*(5939), 151–155.

5. Farrall, M.; Law, H. Y.; Rodeck, C. H.; Warren, R.; Stanier, P.; Super, M.; Lissens, W.; Scambler, P.; Watson, E.; Wainwright, B. et al. First-trimester Prenatal Diagnosis of Cystic Fibrosis with Linked DNA Probes. *Lancet* **1986,** *1*(8495), 1402–1405.

6. Orkin, S. H.; Kazazian, H. H Jr., Antonarakis, S. E.; Goff, S. C.; Boehm, C. D.; Sexton, J. P.; Waber, P. G.; Giardina, P. J. Linkage of Beta-thalassaemia Mutations and Beta-globin Gene Polymorphisms with DNA Polymorphisms in Human Beta-globin Gene Cluster. *Nature* **1982,** *296*(5858), 627–631.

7. Saiki, R. K.; Scharf, S.; Faloona, F.; Mullis, K. B.; Horn, G. T.; Erlich, H. A.; Arnheim, N. Enzymatic Amplification of Beta-globin Genomic Sequences and Restriction Site Analysis for Diagnosis of Sickle Cell Anemia. *Science* **1985,** *230*(4732), 1350–1354.

8. Conner, B. J.; Reyes, A. A.; Morin, C.; Itakura, K.; Teplitz, R. L.; Wallace R. B. (1983) Detection of Sickle Cell Beta S-globin Allele by Hybridization with Synthetic Oligonucleotides. *Proc Natl Acad Sci USA* **1983,** *80*(1), 278–282.

9. Orkin, S. H.; Markham, A. F.; Kazazian, H. H. Jr Direct Detection of the Common Mediterranean Beta-thalassemia Gene with Synthetic DNA Probes. An Alternative Approach for Prenatal Diagnosis. **1983,** *J. Clin. Invest.* 71(3), 775–779.

10. Saiki, R. K.; Bugawan, T. L.; Horn, G. T.; Mullis, K. B.; Erlich, H. A. Analysis of Enzymatically Amplified Beta-globin and HLA-DQ Alpha DNA with Allele-specifi c Oligonucleotide Probes. *Nature* **1986,** *324*(6093), 163–166. doi: 10.1038/324163a0 2 Clinical Molecular Diagnostic Techniques: A Brief Review 34

11. Newton, C. R.; Graham, A.; Heptinstall, L. E.; Powell, S. J.; Summers, C.; Kalsheker, N.; Smith, J. C.; Markham, A. F. Analysis of Any Point Mutation in DNA. The Amplification Refractory Mutation System (ARMS). *Nucleic Acids Res.* **1989,** *17*(7), 2503–2516.

12. Venegas, V.; Halberg, M. C. (2012) Quantifi Cation of mtDNA Mutation Heteroplasmy (ARMS qPCR). *Methods Mol. Biol.* 837, 313–326. doi: 10.1007/978-1-61779-504-6_21.

13. Jarvius, J.; Nilsson, M.; Landegren, U. Oligonucleotide Ligation Assay. *Methods Mol. Biol.* **2003,** *212*, 215–228.

14. Schwartz, K. M.; Pike-Buchanan, L. L.; Muralidharan, K.; Redman, J. B.; Wilson, J. A.; Jarvis, M.; Cura, M. G.; Pratt, V. M. Identifi Cation of Cystic Fibrosis Variants by Polymerase Chain Reaction/oligonucleotide Ligation Assay. *J. Mol. Diagn* **2009,** *11*(3), 211–215. doi: S1525-1578(10)60230-9 [pii] 10.2353/jmoldx.2009.080106

15. Bathum, L.; Hansen, T. S.; Horder, M.; Brosen, K. (1998) A Dual Label Oligonucle-otide Ligation Assay for Detection of the CYP2C19*1, CYP2C19*2, and CYP2C19*3 Alleles Involving Time Resolved Fluorometry. *Ther. Drug. Monit.* **1998,** *20*(1), 1–6.

16. Chakravarty, A.; Hansen, T. S.; Horder, M.; Kristensen, S. R. A Fast and Robust Dual-label Nonradioactive Oligonucleotide Ligation Assay for Detection of Factor V Leiden. *Thromb. Haemost.* **1997,** *78*(4), 1234–1236.

17. Nyren, P.; Lundin, A. Enzymatic Method for Continuous Monitoring of Inorganic Pyrophosphate Synthesis. *Anal Biochem.* **1985,** *151*(2), 504–509.

18. Soderback, E.; Zackrisson, A. L.; Lindblom, B.; Alderborn, A. Determination of CYP2D6 Gene Copy Number by Pyrosequencing. *Clin. Chem.* **2005,** *51*(3), 522–531. doi: clinchem.2004.043182 [pii] 10.1373/clinchem.2004.043182

19. Rose, C. M.; Marsh, S.; Ameyaw, M. M.; McLeod, H. L. Pharmacogenetic Analysis of Clinically Relevant Genetic Polymorphisms. *Methods Mol. Med.* **2003,** *85*, 225–237. doi: 10.1385/1-59259-380-1:225

20. Higuchi, R.; Fockler, C.; Dollinger, G.; Watson, R. Kinetic PCR Analysis: Real-time Monitoring of DNA Amplification Reactions. *Biotechnology* (N Y) **1993,** *11*(9), 1026–1030.

21. Gibson, U. E.; Heid, C. A.; Williams, P. M. A Novel Method for Real Time Quantitative RT-PCR. *Genome Res.* **1996,** *6*(10), 995–1001.

22. Heid, C. A.; Stevens, J.; Livak, K. J.; Williams, P. M. Real Time Quantitative PCR. *Genome Res.* **1996,** *6*(10), 986–994.

23. Myers, R. M.; Lumelsky, N.; Lerman, L. S.; Maniatis, T. (1985) Detection of Single Base Substitutions in Total Genomic DNA. *Nature 313*(6002), 495–498.

24. Yoshino, K.; Nishigaki, K.; Husimi, Y. Temperature Sweep Gel Electrophoresis: A Simple Method to Detect Point Mutations. *Nucleic Acids Res.* **1991,** *19*(11), 3153.

25. Chen, T. J.; Boles, R. G.; Wong, L. J. Detection of Mitochondrial DNA Mutations by Temporal Temperature Gradient Gel Electrophoresis. *Clin. Chem.* **1999,** *45*(8 Pt 1), 1162–1167.

26. Alper, O. M.; Wong, L. J.; Young, S.; Pearl, M.; Graham, S.; Sherwin, J.; Nussbaum, E.; Nielson, D.; Platzker, A.; Davies, Z.; Lieberthal, A.; Chin, T.; Shay, G.; Hardy, K.; Kharrazi, M. Identification of Novel and Rare Mutations in California Hispanic and African American Cystic Fibrosis Patients. *Hum. Mutat.* **2004,** *24*(4), 353. doi: 10.1002/humu.9281

27. Tan, D. J.; Bai, R. K.; Wong, L. J. Comprehensive Scanning of Somatic Mitochondrial DNA Mutations in Breast Cancer. *Cancer Res.* **2002,** *62*(4), 972–976.

28. Orita, M.; Iwahana, H.; Kanazawa, H.; Hayashi, K.; Sekiya, T. (1989) Detection of Polymorphisms of Human DNA by Gel Electrophoresis as Single-strand Conformation Polymorphisms. *Proc. Natl. Acad. Sci. USA* 86(8), 2766–2770.

29. White, M. B.; Carvalho, M.; Derse, D.; O'Brien, S. J.; Dean, M. (1992) Detecting Single Base Substitutions as Heteroduplex Polymorphisms. *Genomics 12*(2), 301–306. doi: 0888-7543(92)90377-5 [pii] M.L. Landsverk and L.-J.C. Wong

30. Makino, R.; Yazyu, H.; Kishimoto, Y.; Sekiya, T.; Hayashi, K. F-SSCP: Fluorescence-based Polymerase Chain Reaction-single-strand Conformation Polymorphism (PCR-SSCP) Analysis. *PCR Methods Appl.* **1992,** *2*(1), 10–13.

31. Wang, Y. H.; Barker, P.; Griffith, J. Visualization of Diagnostic Heteroduplex DNAs from Cystic Fibrosis Deletion Heterozygotes Provides an Estimate of the Kinking of DNA by Bulged Bases. *J. Biol. Chem.* **1992,** *267*(7), 4911–4915.

32. Suzuki, Y.; Orita, M.; Shiraishi, M.; Hayashi, K.; Sekiya, T. Detection of Ras Gene Mutations in Human Lung Cancers by single-strand conformation polymorphism analysis of polymerase Chain Reaction Products. *Oncogene* **1990,** *5*(7), 1037–1043.

33. Dockhorn-Dworniczak, B.; Dworniczak, B.; Brommelkamp, L, Bulles J, Horst J, Bocker WW Non-isotopic Detection of Single Strand Conformation Polymorphism

(PCR-SSCP): A Rapid and Sensitive Technique in Diagnosis of Phenylketonuria. *Nucleic Acids Res.* **1991,** *19*(9) 2500.

34. Hogg, A.; Onadim, Z.; Baird, P. N.; Cowell, J. K. Detection of Heterozygous Mutations in the RB1 Gene in Retinoblastoma Patients Using Single-strand Conformation Polymorphism Analysis and Polymerase Chain Reaction Sequencing. *Oncogene* **1992,** *7*(7), 1445–1451.

35. Underhill, P. A.; Jin, L.; Lin, A. A.; Mehdi, S. Q.; Jenkins, T.; Vollrath, D.; Davis, R. W.; Cavalli-Sforza, L. L.; Oefner, P. J. Detection of Numerous Y Chromosome Biallelic Polymorphisms by Denaturing High-performance Liquid Chromatography. *Genome Res.* **1997,** *7*(10), 996–1005.

36. Den Dunnen, J. T.; Van Ommen, G. J. The Protein Truncation Test: A Review. *Hum. Mutat.* **1999,** *14*(2), 95–102. doi: 10.1002/(SICI)1098-1004(1999)14: 2<95:: AID-HUMU1>3.0.CO;2-G [pii] 10.1002/(SICI)1098-1004(1999)14:2<95::AID-HUMU1>3.0.CO;2-G

37. Roest, P. A.; Roberts, R. G.; van der Tuijn, A. C.; Heikoop, J. C.; van Ommen, G. J.; den Dunnen, J. T. (1993) Protein Truncation Test (PTT) to Rapidly Screen the DMD Gene for Translation Terminating Mutations. *Neuromuscul. Disord.* 3(5–6), 391–394. doi: 0960-8966(93)90083-V [pii]

38. Friedl, W.; Aretz, S. (2005) Familial Adenomatous Polyposis: Experience from a Study of 1164 Unrelated German Polyposis Patients. *Hered Cancer Clin. Pract.* 3(3), 95–114. doi: 1897-4287- 3- 3-95 [pii] 10.1186/1897-4287-3-3-95

39. Hogervorst, F. B.; Cornelis, R. S.; Bout, M.; van Vliet M.; Oosterwijk, J. C.; Olmer, R., Bakker, B.; Klijn, J. G.; Vasen, H. F.; Meijers-Heijboer, H., et al. Rapid Detection of BRCA1 Mutations by the Protein Truncation Test. *Nat. Genet.* **1995,** *10*(2), 208–212. doi: 10.1038/ng0695-208

40. Sanger, F.; Nicklen, S.; Coulson, A. R. DNA Sequencing with Chain-terminating Inhibitors. *Proc. Natl. Acad. Sci. USA* **1977,** *74*(12), 5463–5467.

41. Filippi, G.; Rinaldi, A.; Archidiacono, N.; Rocchi, M.; Balazs, I.; Siniscalco, M. Brief Report: Linkage between G6PD and Fragile-X Syndrome. *Am. J. Med. Genet.* **1983,** *15*(1), 113–119. doi: 10.1002/ ajmg.1320150115

42. Mulligan, L. M.; Phillips, M. A.; Forster-Gibson, C. J.; Beckett, J., Partington, M. W.; Simpson, N. E.; Holden, J. J.; White, B. N. (1985) Genetic Mapping of DNA Segments Relative to the Locus for the Fragile-X Syndrome at Xq27.3. *Am. J. Hum. Genet.* **1985,** *37*(3), 463–472.

43. Oberle, I.; Rousseau, F.; Heitz, D.; Kretz, C.; Devys, D.; Hanauer, A.; Boue, J.; Bertheas, M. F.; Mandel, J. L. Instability of a 550-base Pair DNA Segment and Abnormal Methylation in Fragile X Syndrome. *Science* **1991,** *252*(5010), 1097–1102.

44. Richards, R. I.; Holman, K.; Kozman, H.; Kremer, E.; Lynch, M.; Pritchard, M.; Yu, S.; Mulley, J.; Sutherland, G. R. (1991) Fragile X Syndrome: Genetic Localisation by Linkage Mapping of Two Microsatellite Repeats FRAXAC1 and FRAXAC2 which Immediately Flank the Fragile Site. *J. Med. Genet.* 28(12), 818–823.

45. Yu, S.; Pritchard, M.; Kremer, E.; Lynch, M.; Nancarrow, J.; Baker, E.; Holman, K.; Mulley, J.; Warren, S.; Schlessinger, D., et al. Fragile X Genotype Characterized by an Unstable Region of DNA. *Science* **1991,** *252*(5009), 1179–1181. doi: 252/5009/1179 [pii] 10.1126/ science.252.5009.1179

46. Schouten, J. P.; McElgunn, C. J.; Waaijer, R.; Zwijnenburg, D.; Diepvens, F.; Pals, G. (2002) Relative Quantification of 40 Nucleic Acid Sequences by Multiplex Ligation-dependent Probe Amplification. *Nucleic Acids Res. 30*(12), e57.

47. Kallioniemi, A.; Kallioniemi, O. P.; Sudar, D.; Rutovitz, D.; Gray, J. W.; Waldman, F.; Pinkel, D. Comparative Genomic Hybridization for Molecular Cytogenetic Analysis of Solid Tumors. *Science* **1992,** *258*(5083), 818–821. 2 Clinical Molecular Diagnostic Techniques: A Brief Review 36

48. Landsverk, M. L.; Wang, J.; Schmitt, E. S.; Pursley, A. N.; Wong, L. J. Utilization of Targeted Array Comparative Genomic Hybridization, MitoMet, in Prenatal Diagnosis of Metabolic Disorders. *Mol. Genet. Metab.* **2011,** *103*(2), 148–152. doi: S1096-7192(11)00064-3 [pii] 10.1016/j. ymgme.2011.03.003

49. Wang, J.; Zhan, H.; Li, F. Y.; Pursley, A. N.; Schmitt, E. S.; Wong, L. J. Targeted Array CGH as a Valuable Molecular Diagnostic Approach: Experience in the Diagnosis of Mitochondrial and Metabolic Disorders. *Mol. Genet. Metab.* **2012,** *106*(2), 221 -230. doi: S1096-7192(12)00106-0 [pii] 51 10.1016/j.ymgme.2012.03.005

50. Bejjani, B. A.; Saleki, R.; Ballif, B. C.; Rorem, E. A.; Sundin, K.; Theisen, A.; Kashork, C. D.; Shaffer, L. G. Use of Targeted Array-Based CGH for the Clinical Diagnosis of Chromosomal Imbalance: Is Less More? *Am. J. Med. Genet. A* **2005,** *134*(3), 259–267. doi: 10.1002/ajmg.a.30621

51. Stankiewicz, P.; Beaudet, A. L. Use of Array CGH in the Evaluation of Dysmorphology, Malformations, Developmental Delay, and Idiopathic Mental Retardation. *Curr. Opin. Genet. Dev.* **2007,** *17*(3):182–192. doi: S0959-437X(07)00074-3 [pii] 10.1016/j.gde.2007.04.009

52. Wang, D. G.; Fan, J. B.; Siao, C. J.; Berno, A.; Young, P., Sapolsky, R.; Ghandour, G.; Perkins, N.; Winchester, E.; Spencer, J.; Kruglyak, L.; Stein, L.; Hsie, L.; Topaloglou, T.; Hubbell, E.; Robinson, E.; Mittmann, M.; Morris, M. S.; Shen, N.; Kilburn, D.; Rioux, J.; Nusbaum, C.; Rozen, S.; Hudson, T. J.; Lipshutz, R.; Chee, M.; Lander, E. S. Large-Scale Identification, Mapping, and Genotyping of Single Nucleotide Polymorphisms in the Human Genome. *Science* **1998,** *280*(5366), 1077–1082.

53. Landsverk, M. L.; Douglas, G. V.; Tang, S.; Zhang, V. W.; Wang, G. L.; Wang, J.; Wong, L. J. Diagnostic Approaches to Apparent Homozygosity. **2012,** *Genet. Med.* doi: 10.1038/gim.2012.58 gim201258 [pii]

INDEX

S

T

V

Printed in the United States
by Baker & Taylor Publisher Services